Unmanned Robotic Systems and Applications

Edited by Mahmut Reyhanoglu and Geert De Cubber

Published in London, United Kingdom

IntechOpen

Supporting open minds since 2005

Unmanned Robotic Systems and Applications
http://dx.doi.org/10.5772/intechopen.77608
Edited by Mahmut Reyhanoglu and Geert De Cubber

Assistant to the Editor(s) Daniela Doroftei

Contributors
Alexander Ferrein, Stefan Schiffer, Ingrid Scholl, Tobias Neumann, Kai Krückel, Branko Ristic, Chris Gilliam, Abdulkader Joukhadar, Laxmidhar Behera, Archit Krishna Kamath, Vibhu Kumar Tripathi, Atsushi Kakogawa, Shugen Ma

First published in London, United Kingdom, 2020 by IntechOpen
IntechOpen is the global imprint of INTECHOPEN LIMITED, registered in England and Wales, registration number: 11086078, 7th floor, 10 Lower Thames Street, London, EC3R 6AF, United Kingdom
Printed in Croatia

British Library Cataloguing-in-Publication Data
A catalogue record for this book is available from the British Library

Additional hard and PDF copies can be obtained from orders@intechopen.com

Unmanned Robotic Systems and Applications
Edited by Mahmut Reyhanoglu and Geert De Cubber
p. cm.
Print ISBN 978-1-78984-566-2
Online ISBN 978-1-78984-567-9
eBook (PDF) ISBN 978-1-83880-106-9

For EU product safety concerns:
IN TECH d.o.o., Prolaz Marije Krucifikse Kozulić 3, 51000 Rijeka, Croatia, info@intechopen.com or visit our website at intechopen.com.

Meet the editors

Mahmut Reyhanoglu is presently the Chair and Glaxo Wellcome Distinguished Professor of Engineering at the University of North Carolina at Asheville, North Carolina, USA. His extensive research makes use of advanced mathematical techniques and models that arise from fundamental physical principles. His major research interests are in the areas of nonlinear dynamical systems and control theory, with particular emphasis on applications to mechatronics and aerospace systems. He has authored/co-authored several book chapters and over 140 peer-reviewed journal articles/proceedings papers. He served on the IEEE Transactions on Automatic Control Editorial Board and on the IEEE Control Systems Society Conference Editorial Board as an Associate Editor. He also served as international program committee member for several conferences and as a member of AIAA Guidance, Navigation, and Control Technical Committee.

Geert De Cubber is the team leader of the Robotics & Autonomous Systems Unit of the Department of Mechanics of the Belgian Royal Military Academy. He is also a senior researcher at this institute with a research focus on developing robotic solutions for solving security challenges like crisis management, the fight against crime and terrorism, and border security. He received his diploma in Mechanical Engineering in 2001 from the Vrije Universiteit Brussel and his Doctoral Degree in Engineering in 2010 from the Vrije Universiteit Brussel and the Belgian Royal Military Academy. He is and was the coordinator of multiple European and national research projects, like FP7-ICARUS (on the development of search and rescue robots) and H2020-SafeShore (on the development of a threat detection system).

Contents

Preface

The area of unmanned robotic systems is one of the fastest growing industries and has a number of evolving applications. Autonomous robots are ideal candidates for applications such as rescue missions, especially in areas that are difficult to access. Swarm robotics (multiple robots working together) is another exciting application of the unmanned robotics systems, for example, coordinated search by an interconnected group of moving robots to find a source of hazardous emissions. These robots can behave like individuals working in a group without a centralized control. Researchers have developed intelligent control algorithms for the swarms after deep study of animal behavior in herds, bird flocks, and fish schools.

In the field of robotics, the use of Unmanned Aerial Vehicles (UAVs), more commonly known as drones, has drastically increased over the recent years. In particular, there is a special surge of interest in quadrotor drones because of their advantages over fixed-wing drones in terms of their maneuverability and versatility. Moreover, quadrotor drones have a straightforward mechanical design, are relatively cheap to purchase, and are small in size. Quadrotors are widely used in military and civilian applications involving search and rescue, area mapping, surveillance, wildlife protection, and infrastructure inspection. All these applications involve visual mapping of large areas. The popularity of UAVs to perform these tasks is supported by increased technological possibilities in the area of image recognition. UAVs are strongly coupled, inherently nonlinear systems that require advanced nonlinear control techniques.

Strategies employed for the control of UAVs include linearization-based control techniques such as PID and LQR, and nonlinear control methods. The search for increased performance has always been the main goal for control engineers. As multirotor UAVs are highly nonlinear systems, simple control strategies may not suffice for the performance demand. Due to the high degree of nonlinearity, parameter identification can be difficult, which raises uncertainties in the system model. To cope with these model uncertainties, robustness of the control law becomes a necessity. Sliding mode control is a nonlinear control tool and is known to be a robust control technique. To achieve reliable quadcopter control over a wider operational envelope, several nonlinear control methods have recently been developed. Popular nonlinear control methods for quadrotor systems include backstepping, feedback linearization, dynamic inversion, adaptive control, Lyapunov-based robust control, passivity-based control, fuzzy-model approach, and sliding mode control.

This book presents recent studies of unmanned robotic systems and their applications. With its five chapters, the book brings together important contributions from renowned international researchers. Chapter 1 emphasizes robotic search and rescue via in-pipe inspection robots. It gives an overview of a screw-drive in-pipe mobile robot, a three-module parallel arrangement type in-pipe mobile robot, and several types of multi-link articulated wheeled-type in-pipe robots. Chapter 2 is devoted to the problem of autonomous coordinated search by an interconnected group of moving robots for the purpose of find a source of hazardous emissions such as hazardous gas and particles. The chapter introduces a search strategy that operates in a completely decentralized manner, as long as the communication network of the moving robots forms a connected graph. Chapter 3 proposes a vision-based sliding mode control algorithm for autonomous landing of a quadrotor UAV. The effectiveness of the control algorithm is illustrated through experimental results obtained using a DJI Matrice M100 drone. Chapter 4 presents a custom-built 3D laser range platform SWAP and compares it against an architectural laser scanner. The main advantage of the platform is its ability to scan in a continuous mode. The chapter introduces a new mapping tool (mapit) that can support and automate the registration of large sets of point clouds. Finally, Chapter 5 summarizes different advanced control techniques for UAV control. These techniques include backstepping, feedback linearization, and sliding mode control. A commonly known UAV nonlinear model is presented and the proposed control strategies have been implemented using MATLAB. Simulation results are included to demonstrate the effectiveness of these control techniques.

Mahmut Reyhanoglu, Ph.D.
University of North Carolina Asheville,
Mechatronics Engineering Laboratory,
Asheville, North Carolina,
USA

Geert De Cubber
Royal Military Academy of Belgium,
Department of Mechanics, Robotics & Autonomous Systems unit,
Brussels, Belgium

Chapter 1

Robotic Search and Rescue through In-Pipe Movement

Atsushi Kakogawa and Shugen Ma

Abstract

So far, we have been engaged in the research and development of various kinds of robots that could be applied to in-pipe inspections that existing methods (screw-drive type, parallel multi-modular type, and articulated wheeled type) cannot perform. In this chapter, we categorized each in-pipe inspection robot depending on its configuration and structure, which includes the design of the propulsive mechanism, steering mechanism, stretching mechanism, and the locations of the wheel and joint axes. On the basis of this classification and from a developer's point of view, we also discussed the various kinds of robots that we have developed, along with their advantages and disadvantages.

Keywords: robotic inspection, mechanical design, robots used in limited space, mobile robots, image processing

1. Background

The progressive deterioration of aging social infrastructures in urban areas around the world has led to the occurrences of serious accidents one after another. Risks of accidents are mainly hidden, especially in aging bridges, pipelines, ports, and airports, to name a few, all over the country. In particular, water and gas pipe bursts and leaks, explosion, and fire accidents at complexes are growing into a serious problem. A piping accident, for instance, not only cuts the lifeline but also is associated with potential ignition of leaked gas, which necessitates urgent repair and replacement of deteriorated parts. In the process of repairing and replacing pipelines, the most important issues include how to prioritize the repairing place, how to efficiently identify the deteriorated parts in advance, and how to perform the work with minimum necessary cost and personnel.

The common method of inspection practiced up to the present is manual wall thickness measurement from outside of pipes using ultrasonic and magnetic equipment. Practical-wise, such approach consumes time and could be difficult to employ when reaching pipes installed at high places or underground. In addition, some pipelines contain toxic/explosive carbon monoxide (CO) and silane and combustible/flammable gases, which may cause a health hazard to inspection workers. These setbacks suggest the need for cost and effort reduction in maintaining and managing pipelines and in securing safety. Under these circumstances, the recent development of mechanical and electronic technologies, robotic nondestructive inspection technology (NDT) with cameras, and thickness measurement sensors (ultrasonic and magnetic methods) are receiving attention.

So far, a number of methods called smart pipe inspection gage (PIG) have been reported to utilize fluid force in the pipe to push out and move the camera or inspection device. Owing to this passive movement, a route cannot be selected at the branch sections and cannot propel unless the internal pressure of the pipeline is sufficient. In the pipeline business, PIG is not suited to "unpiggable pipelines." Instead, industrial endoscopes with a camera attached to the tip are widely employed. Nevertheless, as the endoscopes require being pressed in with hands, they are not suitable for inspection in long winding pipelines.

To solve this problem, companies, universities, and research institutions have been working on a large number of self-mobile in-pipe inspection robots. The robot's movement can be roughly classified into legged type [1], peristaltic type [2], serpentine type [3], and infinite rotation type [4–10]. The legged-type robot walks in pipes while extending its legs against the inner wall. However, multiple degrees of freedom cause complicated control systems and an increase in the entire robot size. The peristaltic-type robot produces propagating contractive waves found in earthworms and leeches to move as it pushes out its multiple segments in order. Any of the segments always comes in contact with the inner wall of the pipe to support the body; thus, it can move upward at vertical sections. The serpentine type moves in pipes by sending a waveform to an elongated structure consisting of multiple segments as seen in snakes. Unlike conventional planar snake-like robots with passive rollers at their bottom, the directions of the wave and the travel are the same.

Those types are very interesting and important in the sense of scientific investigation on how animal locomotion adapts to tubelike narrow environments. However, the infinite rotation type, such as in drive wheels and crawler mechanisms (belt-driven), was the one substantially studied as it provides a significantly faster and more efficient motion than the abovementioned animal locomotion schemes despite its simple structure and low cost. Thus, this is expected to contribute in checking buildings or infrastructures before and after disasters, especially in entering into a collapsed building through pipes to search for human casualties.

2. Essential mechanisms for in-pipe inspection robots

For each of the in-pipe robots described above, the body structure consists of three essential components: (1) a propulsive mechanism for moving forward and backward, (2) a steering mechanism for turning at bent and branch sections, and (3) an extending mechanism for avoiding slipping and falling at vertical sections. The propulsive and steering mechanisms are very common in the mobile robot field, whereas the extending mechanism is specific to in-pipe mobile robotic applications. A general in-pipe mobile mechanism is shown in **Figure 1**.

We believe that a key point in designing a small and highly adaptable in-pipe robot is its functional complex. If three components (propulsive, steering, and extending) are installed separately, then an increase in size is inevitable. In a sense, the legged-type, peristaltic-type, and serpentine-type locomotion can be regarded as the common principle because the propulsive mechanism works simultaneously as an extension and as a steering component. Moreover, the radial size of the snake and peristaltic robots may be reduced because the robot body itself generates a propulsive force by shifting its body shape, which suggests the nonnecessity of additional motion mechanisms. As mentioned above, animallike locomotion in pipes is slower than wheel-driven locomotion. Therefore, it is important to develop a scheme that combines the advantage of the wheeled mechanism (faster movement) and the snake and peristaltic mechanism (small size).

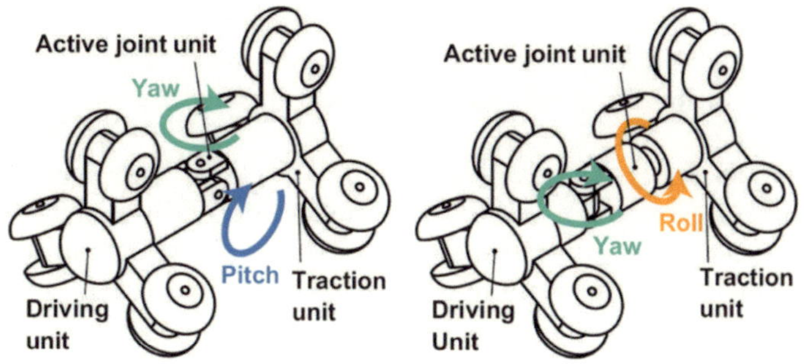

Figure 1.
General in-pipe mobile mechanism.

Structures with several components generally conflict with downsizing. Nonetheless, this issue is solved to some extent by combining multiple functions in one part of the robot (a functional complex). In this study, we tackle two approaches for the functional complex, namely, a differential mechanism and arranging multiple degrees of freedom (DoFs) on a common axis. Conceptually, the differential mechanism approach is applied to a steering mechanism of a screw-drive robot [11] and a step adaptation mechanism of a three-modular robot, whereas the idea of arranging multiple DoFs on a common axis is applied to an articulated wheeled robot.

3. Functional complex by a differential mechanism

An overview of the screw-drive-type in-pipe robot that we first introduced [12] is illustrated in **Figure 2**. This in-pipe robot consists of a front rotator that generates thrust and a rear stator that supports the reaction of the rotator. The rotator has several tilted passive wheels arranged on its circumference and can move forward and backward while tracing a spiral curve.

By arranging a motor and a gear reduction along the pipe axis, the output drive axis can be connected directly to the rotator without changing the direction of rotation through a transmission mechanism, such as a miter. This implies that the screw-drive type can be miniaturized easily, although it would face difficulty passing through T-branches with only a drive mechanism. To solve this challenge, an

Figure 2.
Screw-drive in-pipe mobile robot for 5-in pipelines [12].

active steering joint with a simple miter-geared differential mechanism is installed between the rotator and the stator. The rotator can be swung by only a single actuator in both the longitudinal and lateral directions depending on the in-pipe constraint condition.

Accordingly, the robot can be steered by only a single actuator in both the longitudinal and lateral directions depending on the constraint condition in pipes. Owing to friction, the passive wheels of the middle unit maintain their position during rotation of the steering motor, and the front unit can be swung. Nonetheless, the robot can change its direction of navigation in pipes where steering movement is constrained by the inner wall, e.g., in straight sections. Driven by the orbiting miter gear, the entire middle unit rotates around the central axis; simultaneously, the wheels of the middle unit rotate in the circumferential direction as casters (**Figures 3** and **4**).

Meanwhile, we also developed an in-pipe robot called multi-module parallel arrangement type [13], which has a structure in which multiple belt-driven crawler mechanisms are arranged parallel to the pipe axis and on the circumference. Although it tends to increase in size, a large traction force can be generated by coupling each propulsion force, and the orientation can be changed omnidirectionally by adjusting the speed balance among each module. In this study, we propose a new mechanism called an underactuated parallelogram crawler. We confirm its ability to cope with changes in internal pipe diameter without necessarily an increase in the number of motors (**Figure 5**).

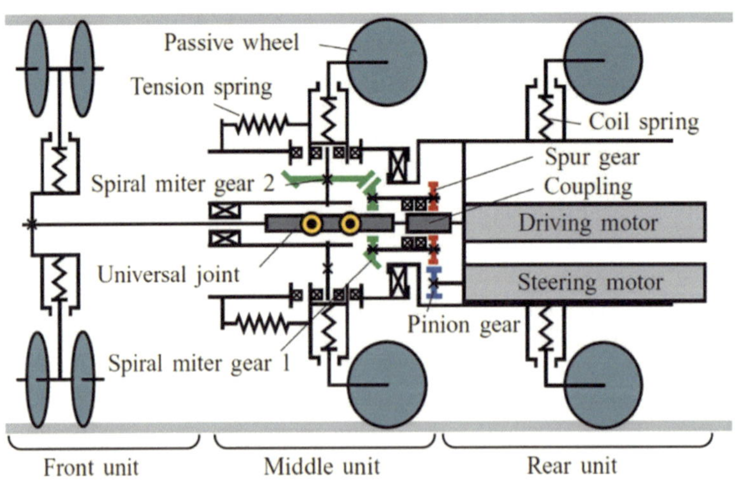

Figure 3.
Schematic of the screw-drive in-pipe mobile robot.

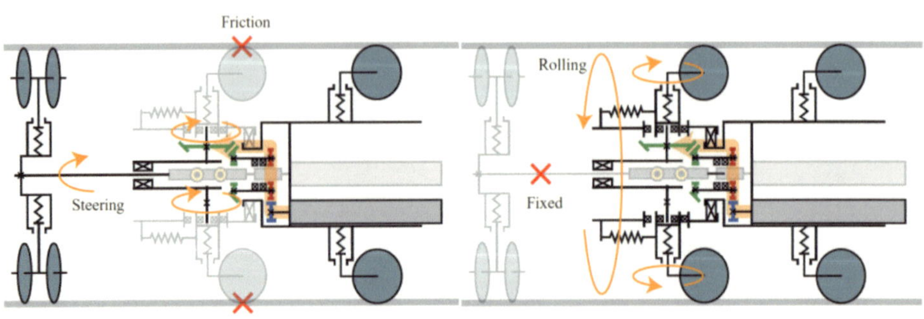

Figure 4.
Steering mode and rolling mode using a miter-geared differential mechanism.

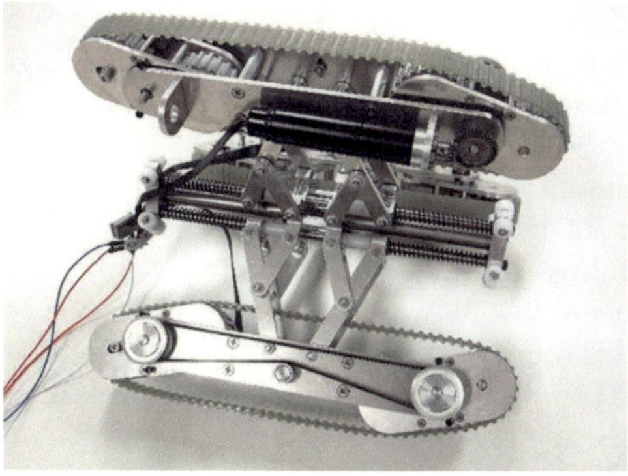

Figure 5.
Three-module parallel arrangement type in-pipe mobile robot for 8-in pipelines [13].

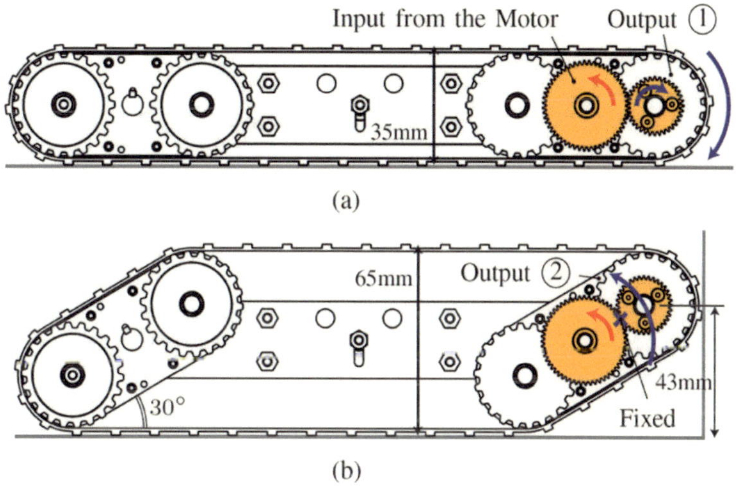

Figure 6.
Driving mode and parallelogram mode using a spur-geared differential mechanism. (a) Driving mode and, (b) Parallelogram mode (arm-lifting).

To achieve differential motion, a pair of spur gears is mounted on the front flipper of each parallelogram crawler module. With the motion of the front flipper constrained by gravity and the pantograph-spring combining expansion mechanism in a normal driving mode, the motor torque is transmitted to the front driving pulley (**Figure 6a**). The front flipper is lifted up once the motion of the robot is stopped (**Figure 6b**). An additional timing belt in these modes enables the simultaneous rotation of the front and rear flippers. To avoid an endless rotation of the flippers, stopper pins are attached to stop at 30°.

4. Functional complex by arranging multiple DoFs on a common axis

On one hand, the screw-drive type can be easily downsized but with an associated limit of travel to pipelines without any junction. On the other hand, the multi-module parallel arrangement type can generate large propulsion by coupling each force but tends to increase in diameter. We thought that the differential mechanism could be one solution for downsizing; however, it leads to the complexity of the whole robot mechanism and eventually causes an increase in size and weight.

As introduced in the earlier sections, the key point for downsizing is combining the three components (propulsive, steering, and extending) in a common component. To achieve this compact design, we have been working on a multi-link-articulated wheeled-type in-pipe robot whose wheel shaft (as propulsive) and joint (as steering and extending) are all arranged on the same axis. This configuration leads to a drastic miniaturization to 3–4 in. in the inner diameter of pipes and is even adaptable to winding pipelines and T-branch [14–19].

An overview of the multi-link-articulated wheeled-type in-pipe robot [20] is shown in **Figure 7**. This robot consists of four links and joints connecting them and moves back and forth using actively rotatable omni wheels installed on each joint axis. A torsional coil spring mounted in each joint allows the robot to form a zigzag shape, making the robot move up in vertical pipes by pressing the omni wheels to the inner wall of pipes. When the robot enters into a bent pipe, the joints can be opened and closed passively according to the shape of the curved section, thus making the robot easily pass through winding pipelines.

Another major feature of this robot is that the rotational axes of all joints are parallel to each other (**Figure 8**). As the positions of all joints move only on the same single plane, the robot cannot pass through bent pipes if the bending direction of the joints does not match the pathway direction of the pipes. However, this is not a disadvantage to the robot. For example, in a situation where the inner wall of pipelines has obstacles, such as holes and dents, the robot can avoid them by displacing the trajectory of the wheels and the obstacle.

To align the bending direction of the robot joints and the pathway direction of the pipe, we proposed a method of changing the robot's orientation around the pipe axis by rolling spherical wheels [21–23] installed at its head and tail ends (**Figure 9**). The spherical wheel rotates freely in the direction of the robot movement; thus, it

Figure 7.
A multi-link-articulated wheeled-type in-pipe robot named AIRo-2.2 [20].

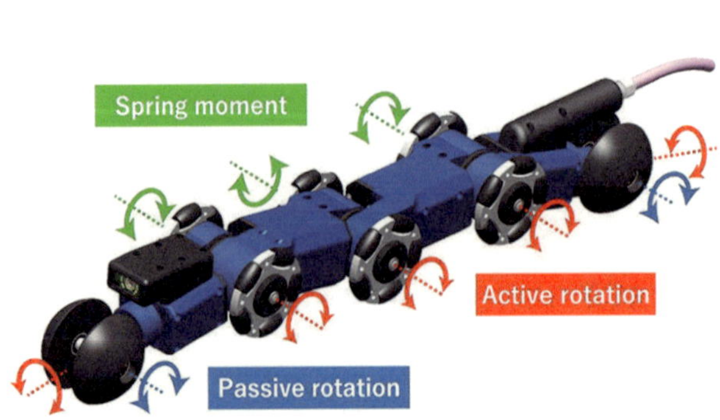

Figure 8.
Two robot orientations depending on the passability to the bent pipe.

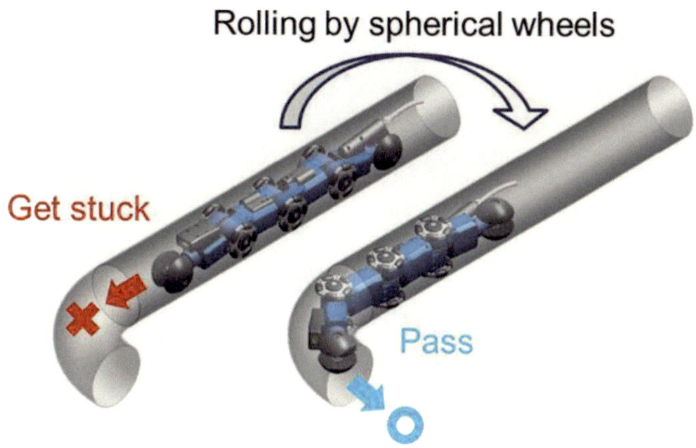

Figure 9.
Experiment in a 15-m pipeline with 12 bent and 1 T-branch pipes.

does not disturb the movement of the omni wheels. Similarly, the omni wheel has several small free rollers arranged on its circumference; thus, they do not interfere with the rolling motion of the robot by the spherical wheels.

We confirmed the robot's ability to pass through a 15-m-length and 4-in-diameter vinyl chloride pipeline, including vertical sections with 12 bent and 1 T-branch pipes, as shown in **Figure 10**. Here, the operator operates the robot using a gamepad while only watching the camera images. It took approximately 6 min for the robot to reach the end of the pipeline and back to the entrance.

We have been in pursuit of a year-by-year improvement of the robot. The multi-link articulated wheeled-type in-pipe robots that we have been developing so far and their extended versions with some modifications (called AIRo-series) are displayed in **Figure 11**.

AIRo-2.2 mini is specially designed for cleaning inside flexible ducts. As it does not have to generate a large traction force, the robot is only composed of two links. By rotating the head brush, the inner surface of the duct can be cleaned. AIRo-3.0 [24, 25] has the same multi-link structure as the AIRo-2.0 series. However, each joint has a differential mechanism to generate two movements: moving back and forth and twisting the body. AIRo-2.4 is a downsized version; from a 4-in diameter, its size was shrunk to 3 in. AIRo-2.3s is the latest version of the in-pipe robot and is equipped with an active joint with both angle and torque control systems (**Figure 11**).

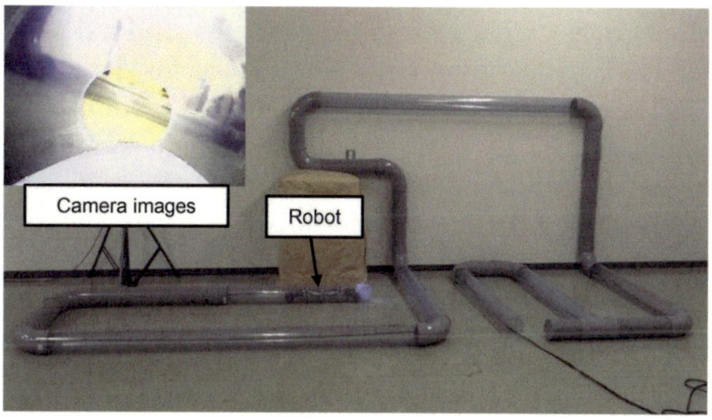

Figure 10.
Active and passive degrees of freedom of the AIRo-2.2 [20].

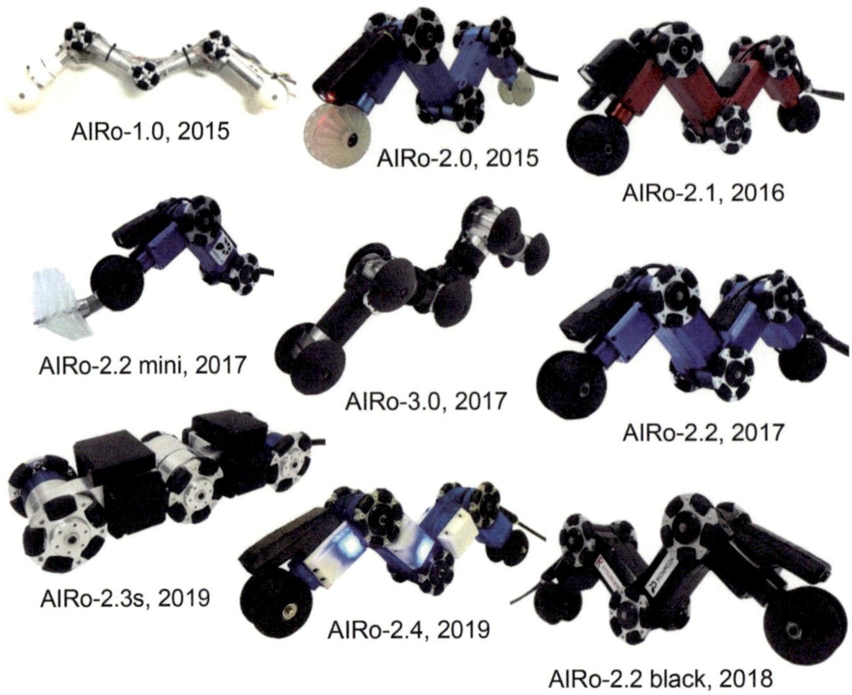

Figure 11.
Multi-link-articulated wheeled-type in-pipe robots in the AIRo-series developed by the authors.

5. Functional complex of camera and operation assistant systems

The robots that we developed do not only assume a straight motion but also roll around the pipe axis through the spherical wheels mounted on the head and tail. However, operators need to select the direction in which the robot orientation should be rotated from only camera images, which normally requires a practiced skill. In addition, this reduces difficulty in operating the robot as it can detect the orientation relative to the pathway direction of the bent pipe by itself.

Regardless of the design, in-pipe inspection robots need at least one illuminator and one camera to view its environment. For this matter, we proposed an anisotropic shadow-based operation assistant method using only a single LED and a camera (**Figure 12**) [26]. By displacing the position of the LED relative to that of the camera, a crescent-shaped shadow appears in the images captured in a bent pipe as illustrated in **Figure 13** [27, 28]. The size, position, and orientation of the shadow depend on the robot's orientation around the pipe axis, and it disappears in a certain robot's orientation (anisotropic shadow). Generally, as for shadow-based navigation systems, shadow disappearance should be avoided to prevent the robot from losing its way. As exclusion, AIRo-2.2 is designed so that it could adapt to a bent pipe without any control when the robot's orientation and the pathway direction of the bent pipe are aligned. By aligning those two specific orientations, the robot can select the optimal orientation to adapt to the bent sections.

Even though the shadow that appears in the bent pipe can be clearly extracted by binarization alone after the threshold is tuned, the shadow that appears in the straight pipe is also detected in straight sections. Therefore, the robot may mistakenly recognize that it is in a bent pipe even if it is in a straight pipe. If the straight pipe and the bent pipe can be distinguished beforehand, then the operation assistant system used to adjust the robot roll orientation can be executed only in bent sections.

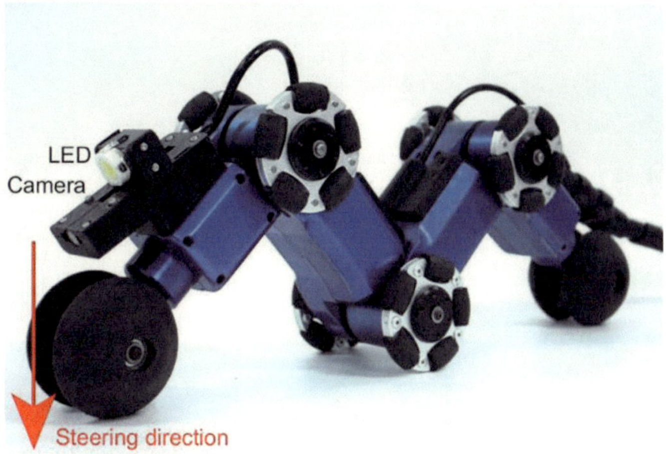

Figure 12.
AIRo-2.2 with a shadow-based operation assistant system [26].

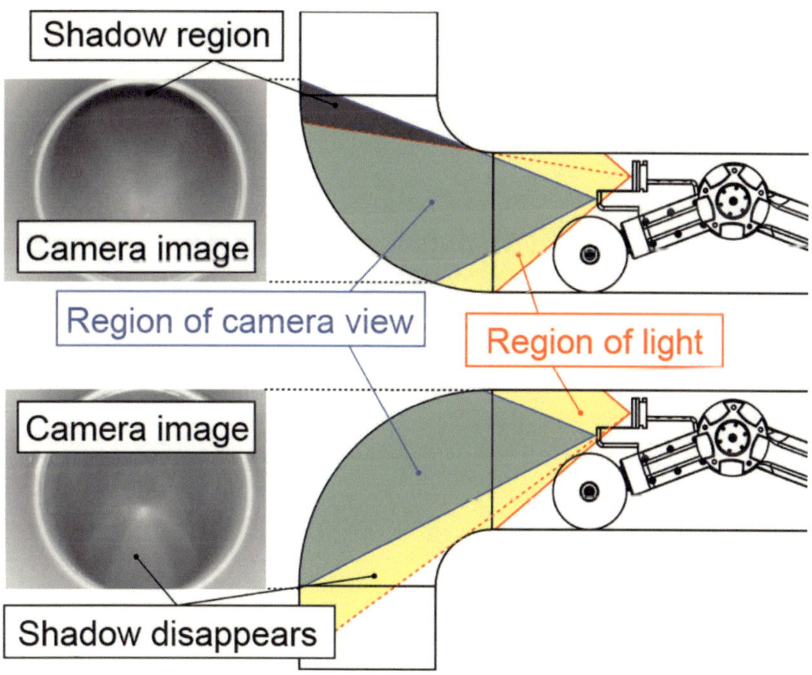

Figure 13.
Principle of crescent shadow appearance in camera images.

In our research, a monochrome image histogram (the relationship between the number of pixels and the brightness value of a camera image) is used to automatically distinguish a bending part and a straight pipe part (**Figure 14**). In straight pipes, the LED brightens the inner pipe wall around the robot. However, the light does not reach the far-off portion of the pipe (the center of the camera image), consequently leading to an even distribution in the luminance values from low to high on the image histogram (**Figure 14a**).

In bent sections, as the LED light is brightly reflected in many areas (**Figure 14b**) in the camera images (the distance between the LED and the wall of the pipe is close), the luminance value is concentrated at the high brightness on the image histogram (**Figure 15**). This difference can be distinguished by the

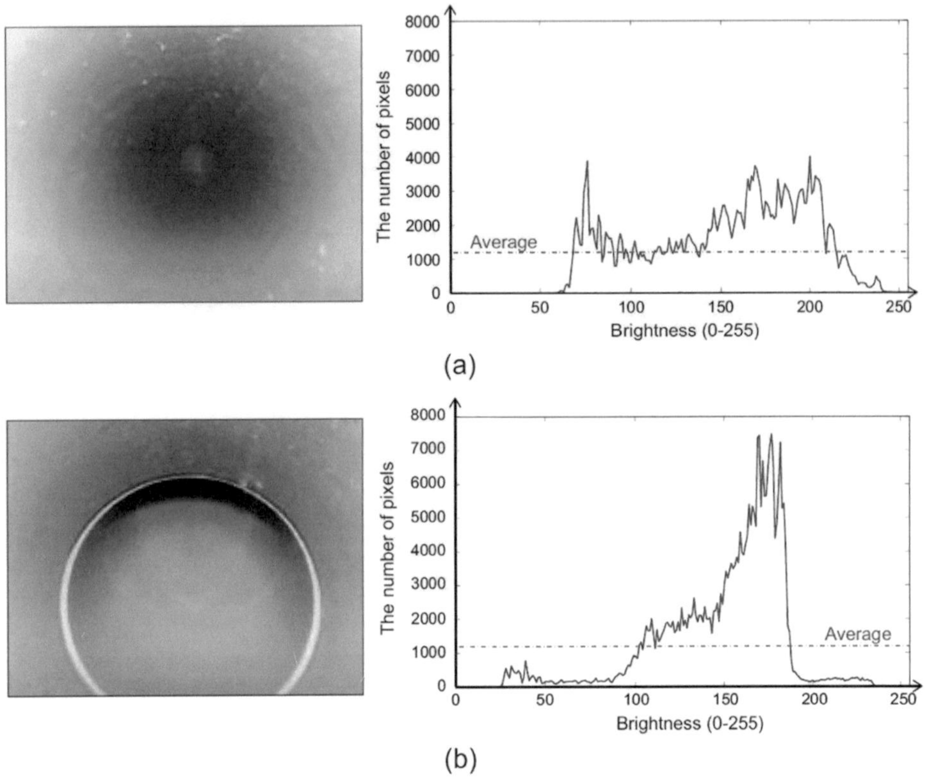

Figure 14.
Image histograms captured in a straight section (a) and a bent section (b).

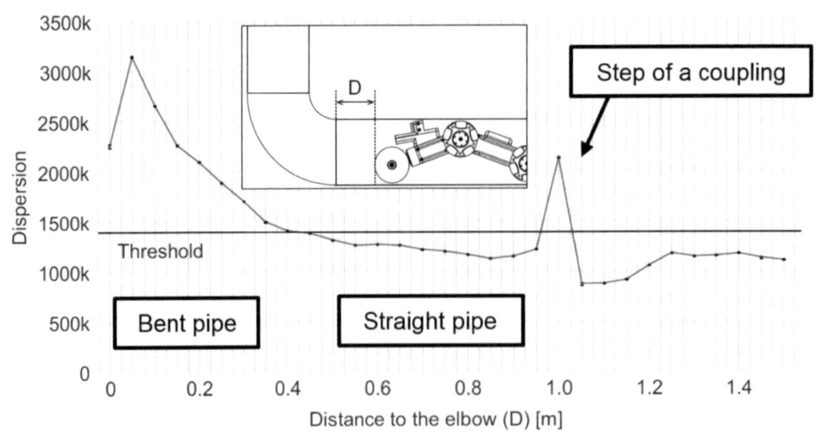

Figure 15.
Variance value depending on the distance between the robot and the entrance of the bent pipe.

calculation of the variance of the image histogram. Apparently, the value of variance increases in bent pipes, and it decreases in straight pipes.

We set the threshold of variance to 1.4 million and found that the robot recognizes the bent pipe when it exceeded about D = 0.4 m. There is a portion where the variance value increases rapidly near D = 1.0 m in the graph mainly because of the influence of reflected light from the step part of the connecting pipe. Although the robot incorrectly recognizes it as a bent pipe in a straight section, traveling is not inhibited even if roll rotation is performed. At D = 0 m, the variance value decreases because the shadow image of the bent pipe is detected.

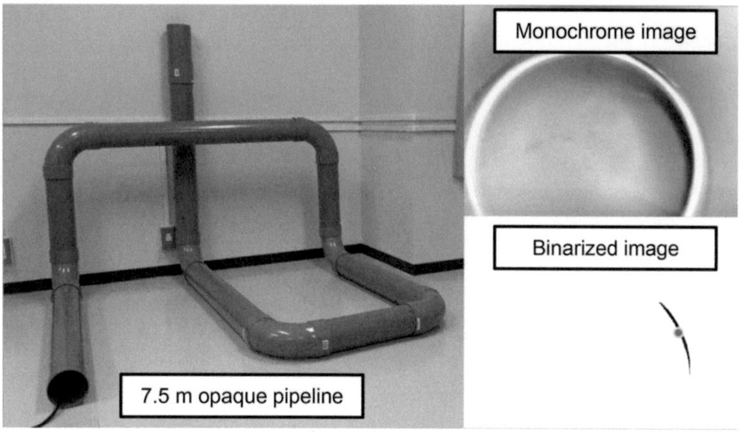

Figure 16.
Experimental result of the robot with the shadow-based operation assistant system.

We confirmed from the experiments that the robot with the shadow-based operation assistant system could travel by approximately 7.5 m in length, including three vertical pipes and seven bent pipes. At this time, the operator used only one button to make the robot move forward or stop (**Figure 16**).

6. Conclusions

Herein, we introduced our researched and developed in-pipe inspection robots and their shadow-based operation assistant system (orientation adjustment). The key point to designing such robot for various pipelines available is downsizing and simplification by the functional complex. Our approach for this functional complex included a differential mechanism, arrangement of multiple DoFs on a common axis, and usage of a camera not only for inspection but also for the operation assistant system. At the present stage, we are testing such approach in simulated pipelines installed in our laboratory. Nonetheless, we will continue to improve the development while collaborating with the user company and are planning to carry out experiments on the actual site.

Acknowledgements

This work was supported by JSPS KAKENHI Grant Number 17K14632, Gesuido Academic Incubation to Advanced Project (GAIA Project) of the Ministry of Land, Infrastructure, Transport and Tourism and many other cooperative companies as commissioned research. We would like to thank all those who supported.

Conflict of interest

There is no conflict of interest regarding the publication of this article.

Author details

Atsushi Kakogawa and Shugen Ma*
Department of Robotics, Ritsumeikan University, Japan

*Address all correspondence to: shugen@se.ritsumei.ac.jp

References

[1] Zagler A, Pfeiffer F. "MORITZ" a pipe crawler for tube junctions. In: Proceeding of the IEEE International Conference on Robotics and Automation; 2003. pp. 2954-2959

[2] Yanagida T, Adachi K, Nakamura T. Development of bellows-type artificial rubber muscle and application to peristaltic crawling endoscopic robot. Journal of Robotics and Mechatronics. 2013;**25**(4):748-754

[3] Kuwada A, Adomi Y, Suzumori K, Kanda T, Wakimoto S, Kadowaki N. Snake-like robot negotiating three-dimensional pipelines. In: Proceeding of the IEEE International Conference on Robotics and Biomimetics; 2007. pp. 989-994

[4] Choi HR, Ryew SM. Robotic system with active steering capability for internal inspection of urban gas pipelines. Mechatronics. 2002;**12**:713-736

[5] Rome E, Hertzberg J, Kirchner F, Licht U, Christaller T. Towards autonomous sewer robots: The MAKRO project. Urban Water. 1999;**1**:57-70

[6] Schempf H, Mutschler E, Gavaert A, Skoptsov G, Crowley W. Visual and nondestructive evaluation inspection of live gas mains using the Explorer™ family of pipe robots. Journal of Field Robotics. 2010;**27**:217-249

[7] Birkenhofer C, Hoffmeister M, Zollner JM, Dillmann R. Compliant motion of a multi-segmented inspection robot. In: Proceeding of the IEEE/RSJ International Conference Intelligent Robots and Systems; 2005. pp. 2632-2637

[8] Pfotzer L, Staehler M, Hermann A, Roennau A, Dillmann R. KAIRO 3: Moving over stairs & unknown obstacles with reconfigurable snake-like robots. In: Proceeding of the European Conference on Mobile Robots; 2015. pp. 1-6

[9] Kim HM, Choi YS, Mun HM, Yang SU, Park CM, Choi HR. 2-2D Differential gear mechanism for robot moving inside pipelines. In: Proceeding of the IEEE/RSJ International Conference on Intelligent Robots and Systems; 2015. pp. 1152-1157

[10] Kim HM, Yang SU, Choi YS, Mun HM, Park CM, Choi HR. Design of back-drivable joint mechanism for in-pipe robot. In: Proceeding of the IEEE/RSJ International Conference on Intelligent Robots and Systems; 2015. pp. 3779-3784

[11] Horodinca M, Doroftei I, Mignon E, Preumont A. A simple architecture for in-pipe inspection robots. In: Proceeding of the International Colloquium on Autonomous and Mobile Systems; 2002. pp. 61-64

[12] Kakogawa A, Nishimura T, Ma S. Designing arm length of a screw drive in-pipe robot for climbing vertically positioned bent pipes. Robotica. 2016;**34**(2):306-327

[13] Kakogawa A, Ma S, Hirose S. An in-pipe robot with underactuated parallelogram crawler modules. In: Proceeding of the IEEE International Conference on Robotics and Automation; 2014. pp. 1687-1692

[14] Dertien E, Stramigioli S, Pulles K. Development of an inspection robot for small diameter gas distribution mains. In: Proceeding of the IEEE International Conference on Robotics and Automation; 2011. pp. 5044-5049

[15] Dertien E, Foumashi M, Pulles K, Stramigioli S. Design of a robot for in-pipe inspection using omnidirectional wheels and active stabilization.

In: Proceeding of the IEEE International Conference on Robotics and Automation; 2014. pp. 5121-5126

[16] Isomura K, Hirose S. Development of articulated spherical wheeled in-pipe robot "ThesV". In: Proceeding of the JSME Conference on Robotics and Mechatronics; 2011. pp. 2A2-M02 (in Japanese)

[17] Debenest P, Guarnieri M, Hirose S. Pipe tron series-robots for pipe inspection. In: Proceeding of the 3rd International Conference on Applied Robotics for the Power Industry; 2014. pp. 1-6

[18] Fjerdingen S, Liljebäck P, Transeth A. A snake-like robot for internal inspection of complex pipe structures (PIKo). In: Proceeding of the IEEE/RSJ International Conference on Intelligent Robots and Systems; 2009. pp. 5665-5671

[19] Vradis GC, Leary W. Development of an inspection platform and a suite of sensors for assessing corrosion an mechanical damage on unpiggable transmission mains. Technical Report of NGA and Foster-Miller; 2004

[20] Kakogawa A, Ma S. Design of a multilink-articulated wheeled pipeline inspection robot using only passive elastic joints. Advanced Robotics. 2018;**32**(1):37-50

[21] Tadakuma K. Tetrahedral mobile robot with novel ball shape wheel. In: Proceeding of the First IEEE/RAS-EMBS International Conference on Biomedical Robotics and Biomechatronics; 2006. pp. 946-952

[22] Tadakuma K, Tadakuma R, Berengeres J. Development of holonomic omnidirectional vehicle with "Omni-Ball": Spherical wheels. In: Proceeding of the IEEE/RSJ International Conference on Intelligent Robots and Systems; 2007. pp. 33-39

[23] Ye C, Ma S. Development of an omnidirectional mobile platform. In: Proceeding of IEEE International Conference on Mechatronics and Automation; 2009. pp. 1111-1115

[24] Kakogawa A, Oka Y, Ma S. Multi-link articulated wheeled in-pipe robot with underactuated twisting joints. In: Proceeding of the IEEE International Conference on Mechatronics and Automation; 2019. pp. 942-947

[25] Oka Y, Kakogawa A, Ma S. Stopper angle design for a multi-link articulated wheeled in-pipe robot with underactuated twisting joints. In: Proceeding of the IEEE/RSJ International Conference on Intelligent Robots and Systems; 2019. pp. 973-978

[26] Kakogawa A, Komurasaki Y, Ma S. Anisotropic shadow-based operation assistant for a pipeline-inspection robot using a single illuminator and camera. In: Proceeding of the IEEE/RSJ International Conference on Intelligent Robots and Systems; 2017. pp. 1305-1310

[27] Lee DH, Moon H, Choi HR. Landmark detection methods for in-pipe robot traveling in urban gas pipelines. Robotica. 2016;**34**(3):601-618

[28] Lee J, Roh S, Kim DW, Moon H, Choi HR. In-pipe robot navigation based on the landmark recognition system using shadow images. In: Proceeding of the IEEE International Conference on Robotics and Automation; 2009. pp. 1857-1862

Chapter 2

Decentralised Scalable Search for a Hazardous Source in Turbulent Conditions

Branko Ristic and Christopher Gilliam

Abstract

The problem is autonomous coordinated search by an interconnected group of moving robots for the purpose of finding and localising a source of hazardous emissions (e.g., gas and particles). Dispersion of the emitted substance is assumed to be affected by turbulence, resulting in the absence of concentration gradients. The chapter proposes a search strategy that operates in a completely decentralised manner, as long as the communication network of the moving robots forms a connected graph. By decentralised operation, we mean that each moving robot is reasoning (i.e., estimating the source location and making decisions on robot motion) locally. Coordination of the group is achieved by consensus via communication with the neighbours only, in a manner which does not require global knowledge of the communication network topology.

Keywords: autonomous search, machine intelligence, sequential Monte Carlo estimation, infotaxis

1. Introduction

Searching strategies for finding targets using appropriate sensing modalities are of great importance in many aspects of life. In the context of national security, there could be a need to find a source of hazardous emissions [1–3]. Similarly, rescue and recovery missions may be tasked with localising a lost piece of equipment that is emitting weak signals [4]. Biological applications include, for example, protein searching for its specific target site on DNA [5], or foraging behaviour of animals in their search for food or a mate [6, 7]. The objective of search research [8] is to develop optimal strategies for localising a target in the shortest time (on average), for a given search volume and sensing characteristics.

The use of autonomous vehicles in dangerous missions, such as finding a source of hazardous emissions, has become widespread [9–11]. Existing approaches to the search and localisation in the context of atmospheric releases can be loosely divided into three categories: up-flow motion methods, concentration gradient-based methods and information gain-based methods, also known as *infotaxis*. Both the up-flow motion methods and the concentration gradient methods are simple, in the sense that they require only a limited level of spatial perception [12]. Their limitations manifest in the presence of turbulent flows, due to the absence of concentration gradients, when the plume typically consists of time-varying disconnected patches. The information gain-based methods [13] have been developed specifically for searching in turbulent flows. In the absence of a smooth distribution of concentration (e.g., due to turbulence), this strategy directs the searching robot(s) towards the highest information gain. As a theoretically principled approach, where the source-parameter estimation is carried out in the Bayesian framework and the searching platform motion control is based on the information-theoretic principles, the infotaxic (or cognitive) search strategies have attracted a great deal of interest [3, 14–23].

This chapter summarizes our recent results in development of an autonomous infotaxic coordinated search strategy for a group of robots, searching for an emitting hazardous source in open terrain under turbulent conditions. The assumption is that the search platforms can move and sense. Two types of sensor measurements are collected sequentially: (a) the concentration of the hazardous substance; (b) the platform location within the search domain. Due to the turbulent transport of the emitted substance, the concentration measurements are typically sporadic and fluctuating. The searching platforms form a moving sensor network, thus enabling the exchange of data and a cooperative behaviour. The multi-robot infotaxis have already been studied in [16, 17, 20, 24]. However, all mentioned references assumed *all-to-all* (i.e., fully connected) communication network with *centralised fusion and control* of the searching group.

We develop an approach where the group of searching robots operate in a fully decentralised coordinated manner. Decentralised operation means that each searching robot performs the computations (i.e., source estimation and path planning) locally and independently of other platforms. Having a common task, however the robotic platforms must perform in a coordinated manner. This coordination is achieved by exchanging the data with immediate neighbours only, in a manner which does not require the global knowledge of the communication network topology. For this reason, the proposed approach is scalable in the sense that the complexities for sensing, communication, and computing per sensor platform are independent of the sensor network size. In addition, because all sensor platforms are treated equally (no leader-follower hierarchy), this approach is robust to the failure of any of the searching agents. The only requirement for avoiding the break-up of the searching formation is that the communication graph of the sensor network remains *connected* at all times. Source-parameter estimation is carried out sequentially, and on each platform independently, using a Rao-Blackwellised particle filter. Platform path planning, in the spirit of *infotaxis*, is based on entropy-reduction and is also carried out independently on every platform.

2. Mathematical models

First, we describe the measurement model. The concentration measurements are modelled using a Lagrange encounters model developed in [13], based on an open field assumption and a two-dimensional geometry. Let ith robotic vehicle position $(i = 1, 2, \ldots, N)$ at time t_k be denoted by $\mathbf{r}_k^i \in \mathbb{R}^2$. Suppose that the emitting source is located at coordinates specified by the vector $\mathbf{r}_0 = [X_0, Y_0]^\intercal$ and its release rate, or strength, is Q_0. The goal of the search is to detect and estimate the source-parameter vector $\boldsymbol{\eta}_0 = \left[\mathbf{r}_0^\intercal \; Q_0\right]^\intercal$ in the shortest possible time. The particles released from the source propagate with combined molecular and turbulent isotropic diffusivity D, but can also be advected by wind. The released particles have an average lifetime τ before being absorbed. Let the *average* wind characteristics be the speed U and direction, which by convention, coincides with the direction of the x axis. Suppose a spherical concentration measuring sensor of small radius a is mounted on the ith robot, whose position at time k is[1] $\mathbf{r}_k^i = [x_k^i, y_k^i]^\top$. This sensor will experience a series of encounters with the particles released from the emitting source. The average rate of encounters can be modelled as follows [13]:

$$R(\boldsymbol{\eta}_0, \mathbf{r}_k^i) = \frac{Q_0}{\ln\left(\frac{\lambda}{a}\right)} \exp\left[\frac{(X_0 - x_k^i)U}{2D}\right] \cdot K_0\left(\frac{d_k^i(\mathbf{r}_0, \mathbf{r}_k^i)}{\lambda}\right) \tag{1}$$

where D, τ and U are known environmental parameters, $d_k^i(\mathbf{r}_0, \mathbf{r}_k^i) = \sqrt{(x_k^i - X_0)^2 + (y_k^i - Y_0)^2}$ is the distance between the source and the ith sensor platform, K_0 is the modified Bessel function of the second kind of order zero, and $\lambda = \sqrt{D\tau/\left(1 + \frac{U^2\tau}{4D}\right)}$ depends on environmental parameters only.

The probability that a sensor at location $\mathbf{r}_k^i$ is hit by $z \in \mathbb{Z}^+ \cup \{0\}$ dispersed particles (where z is a non-negative integer) during a time interval t_0 is Poisson distributed, i.e.,

$$\mathcal{P}(z; \mu_k^i) = \frac{(\mu_k^i)^z}{z!} e^{-\mu_k^i}. \tag{2}$$

Parameter $\mu_k^i = t_0 \cdot R(\boldsymbol{\eta}_0, \mathbf{r}_k^i)$ in (2) is the mean number of particles expected to reach the sensor at location $\mathbf{r}_k^i$ during interval t_0. Eq. (2) expressed the likelihood function of a concentration measurement z_k^i collected by ith sensor, i.e., $\ell(z_k^i | \boldsymbol{\eta}_0) = \mathcal{P}(z_k^i; \mu_k^i)$.

The motion model of a coordinated group of robots is described next. Let the pose vector of the ith robot platform at time t_k be denoted $\boldsymbol{\theta}_k^i = [(\mathbf{r}_k^i)^\top, \ \phi_k^i]^\top$, where $\mathbf{r}_k^i = [x_k^i, y_k^i]^\top$ has already been introduced and ϕ_k^i is the vehicle heading. The group of searching vehicles moves in a formation. The centroid of the formation at time t_k is specified by coordinates:

$$x_k^c = \frac{1}{N}\sum_{i=1}^{N} x_k^i, \quad y_k^c = \frac{1}{N}\sum_{i-1}^{N} y_k^i. \tag{3}$$

For each platform $i = 1, \ldots, N$, the offset $(\Delta x_i, \Delta y_i)$ from the centroid (x_k^c, y_k^c) is predefined and known to it (i.e., $x_k^i = x_k^c + \Delta x_i$, $y_k^i = y_k^c + \Delta y_i$).

The measurements of concentration are taken at time instants t_k, $k = 1, 2, \cdots$. Between two consecutive sensing instants, each platform is moving. Let the duration of this interval (referred to as the *travel time*) for the ith platform be $T_k^i \geq 0$. The assumption is that sensing is suppressed during the travel time.

Motion of the ith platform during interval T_k^i is controlled by linear velocity V_k^i and angular velocity Ω_k^i. Given that the motion control vector $\mathbf{u}_k^i = [V_k^i, \Omega_k^i, T_k^i]^\top$ is applied to the ith platform, its dynamics during a short integration time interval $\delta \ll T_k^i$ can be modelled by a Markov process whose transitional density is $\pi(\boldsymbol{\theta}_t^i | \boldsymbol{\theta}_{t-\delta}^i, \mathbf{u}_k^i) = \mathcal{N}(\boldsymbol{\theta}_t^i; \beta(\boldsymbol{\theta}_{t-\delta}^i, \mathbf{u}_k^i), \mathbf{Q})$. The process noise covariance matrix $\mathbf{Q}$ captures the uncertainty in motion due to the unforeseen disturbances. The vehicle motion function $\beta(\boldsymbol{\theta}_{t-\delta}^i, \mathbf{u}_k^i)$ is:

[1] Robot locations are assumed to be non-coincidental with the source location $\mathbf{r}_0$.

$$\beta\left(\boldsymbol{\theta}_{t-\delta}^i, \mathbf{u}_k^i\right) = \boldsymbol{\theta}_{t-\delta}^i + \delta \begin{bmatrix} V_k^i \cos\left(\phi_{k-1}^i\right) \\ V_k^i \sin\left(\phi_{k-1}^i\right) \\ \Omega_k^i \end{bmatrix} + \mathbf{B}_{k-1}^i, \tag{4}$$

where vector $\mathbf{B}_{k-1}^i = \begin{bmatrix} \varepsilon_x^i \frac{\delta}{T_k^i} & \varepsilon_y^i \frac{\delta}{T_k^i} & 0 \end{bmatrix}^{\mathrm{T}}$ is introduced to compensate for a distortion of the formation due to process noise with parameters:

$$\varepsilon_x^i = \overline{x}_{k-1}^i - \left(x_{k-1}^i - \Delta x_i\right) \tag{4a}$$

$$\varepsilon_y^i = \overline{y}_{k-1}^i - \left(y_{k-1}^i - \Delta x_i\right). \tag{4b}$$

Here, $\overline{x}_{k-1}^i$ and $\overline{y}_{k-1}^i$ are the estimates of the coordinates of the formation centroid at $k-1$ (that is of x_{k-1}^c and y_{k-1}^c, respectively) available to the ith platform. Coordinates x_{k-1}^i and y_{k-1}^i refer to the *known* ith vehicle position at $k-1$. **Figure 1** illustrates the trajectories of $N = 7$ autonomous vehicles in a formation using the described transitional density $\pi\left(\boldsymbol{\theta}_t^i | \boldsymbol{\theta}_{t-\delta}^i, \mathbf{u}_k^i\right)$. In the absence of process noise (i.e., $\mathbf{Q} = 0$), the vehicles would move in a perfect formation if (a) all control vectors are identical (i.e., $\mathbf{u}_k^1 = \mathbf{u}_k^2 = \cdots = \mathbf{u}_k^N$), and (b) all headings are identical (i.e., $\phi_{k-1}^1 = \phi_{k-1}^2 = \cdots = \phi_{k-1}^i$). In this case, each platform would know the true coordinates of the formation centroid (i.e., $\overline{x}_k^i = x_k^c, \overline{y}_k^i = y_k^c$, for $i = 1, \ldots, N$), and hence the correction vectors $\mathbf{B}_{k-1}^i$ would be zero.

A robotic platform can communicate with another platform of the formation, if their mutual distance is smaller than a certain range $R_{\max}$. Because of process noise in motion, the distance between the vehicles in the formation will vary and consequently the topology of the communication network graph may also vary. For simplicity, we will assume that communication links (when established) are error free. **Figure 1** illustrates the communication graphs of a formation consisting of $N = 7$ searching platforms at two consecutive time instants.

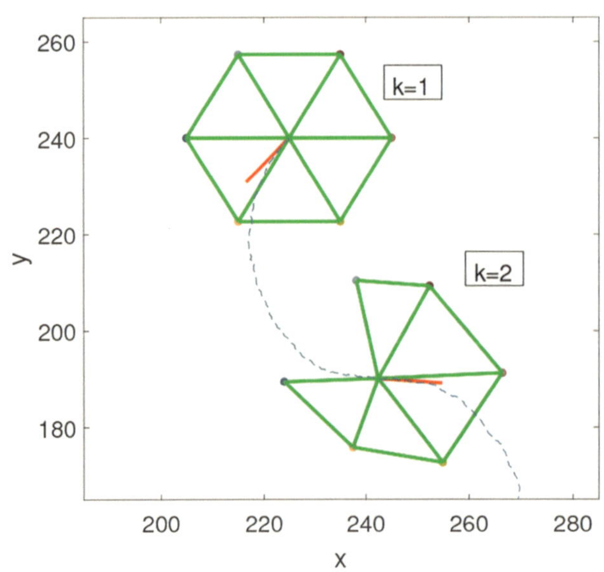

Figure 1.
An example of a formation of N = 7 searching platforms at k = 1, 2. The communication graphs (based on established links between the platforms) are indicated with green lines. Note that communication network topology is time-varying. The red line, starting from the centroid of the formation, indicates the instantaneous velocity vector.

3. Decentralised sequential estimation

Estimation and robot motion control are carried out using the measurement dissemination-based decentralised fusion architecture [25]. Measurement locations[2] and the corresponding measured concentration values, i.e., the triple (x_k^i, y_k^i, z_k^i), are exchanged via the communication network. The protocol is iterative. In the first iteration, platform i broadcasts its triple to its neighbours and receives from them their measurement triples. In the second, third and all subsequent iterations, platform i broadcasts its newly acquired triples to the neighbours, and accepts from them only the triples that this platform has not seen before (newly acquired). Providing that the communication graph is connected, after a sufficient number of iterations (which depends on the topology of the graph), a complete list of measurement triples from all platforms in the formation, denoted $d_k = \{(x_k^i, y_k^i, z_k^i)\}_{1 \le i \le N}$, will be available at each platform.

Suppose the posterior density function of the source at discrete-time $k-1$ and platform i be denoted $p_i(\boldsymbol{\eta}_0|d_{1:k-1})$, where $d_{1:k-1} \equiv d_1, d_2, \cdots, d_{k-1}$. Given $p_i(\boldsymbol{\eta}_0|d_{1:k-1})$ and d_k, the problem of sequential estimation is to compute the posterior at time k, i.e., $p_i(\boldsymbol{\eta}_0|d_{1:k})$. Using the Bayes rule, the posterior is

$$p_i(\boldsymbol{\eta}_0|d_{1:k}) = \frac{g(d_k|\boldsymbol{\eta}_0)p_i(\boldsymbol{\eta}_0|d_{1:k-1})}{\int g(d_k|\boldsymbol{\eta}_0)p_i(\boldsymbol{\eta}_0|d_{1:k-1})d\boldsymbol{\eta}_0} \tag{5}$$

where $g(d_k|\boldsymbol{\eta}_0)$ is the likelihood function. Assuming that individual platform measurements are conditionally independent, $g(d_k|\boldsymbol{\eta}_0)$ can be expressed as

$$g(d_k|\boldsymbol{\eta}_0) = \prod_{i=1}^{N} \ell(z_k^i|\boldsymbol{\eta}_0) = \prod_{i=1}^{N} \mathcal{P}(z_k^i; Q_0\ \rho(\mathbf{r}_0, \mathbf{r}_k^i)) \tag{6}$$

where

$$\rho(\mathbf{r}_0, \mathbf{r}_k^i) = t_0 R(\boldsymbol{\eta}_0, \mathbf{r}_k^i)/Q_0 = \frac{t_0}{\ln\left(\frac{\lambda}{a}\right)} \exp\left[\frac{(X_0 - x_k^i)U}{2D}\right] \cdot K_0\left(\frac{d_k^i(\mathbf{r}_0, \mathbf{r}_k^i)}{\lambda}\right) \tag{7}$$

is independent of Q_0. The posterior density $p_i(\boldsymbol{\eta}_0|d_{1:k})$ is computed using the Rao-Blackwell dimension reduction scheme [26]. Using the chain rule, the posterior can be expressed as:

$$p_i(\boldsymbol{\eta}_0|d_{1:k}) = p_i(Q_0|\mathbf{r}_0, d_{1:k}) \cdot\ p_i(\mathbf{r}_0|d_{1:k}) \tag{8}$$

where the posterior of source strength $p_i(Q_0|\mathbf{r}_0, d_{1:k})$ will be worked out analytically, while the posterior of source position $p_i(\mathbf{r}_0|d_{1:k})$ will be computed using a particle filter. Following [27], we express the posterior $p_i(Q_0|\mathbf{r}_0, d_{1:k-1})$ with the Gamma distribution whose shape and scale parameters are κ_{k-1} and ϑ_{k-1}, respectively. That is

$$\begin{aligned} p_i(Q_0|\mathbf{r}_0, d_{1:k-1}) &= \mathcal{G}(Q_0; \kappa_{k-1}, \vartheta_{k-1}) \\ &= \frac{Q_0^{(\kappa_{k-1}-1)} e^{-Q_0/\vartheta_{k-1}}}{\vartheta_{k-1}^{\kappa_{k-1}} \Gamma(\kappa_{k-1})}. \end{aligned} \tag{9}$$

[2] Because the measurement locations are assumed to be known exactly, they will not be treated as random variables.

Since the conjugate prior of the Poisson distribution is the Gamma distribution [28], the posterior $p(Q_0|\mathbf{r}_0, d_{1:k})$ is also a Gamma distribution with updated parameters κ_k and ϑ_k, i.e., $p(Q_0|\mathbf{r}_0, d_{1:k}) = \mathcal{G}(Q_0; \kappa_k, \vartheta_k)$. The computation of κ_k and ϑ_k can be carried out analytically as a function of $\mathbf{r}_0$ and the measurement set $d_k = \{(\mathbf{r}_k^i, z_k^i)\}_{1 \le i \le N}$ [27]:

$$\kappa_k = \kappa_{k-1} + \sum_{i=1}^{N} z_k^i, \quad \vartheta_k = \frac{\vartheta_{k-1}}{1 + \vartheta_{k-1} \sum_{i=1}^{N} \rho(\mathbf{r}_0, \mathbf{r}_k^i)}. \tag{10}$$

The parameters of the prior for source strength, $p(Q_0) = \mathcal{G}(\kappa_0, \vartheta_0)$ are chosen so that this density covers a large span of possible values of Q_0.

Next, we turn our attention to the posterior of source position $p_i(\mathbf{r}_0|d_{1:k})$ in the factorised form (8). Given $p(\mathbf{r}_0|d_{1:k-1})$, the update step of the particle filter using d_k applies the Bayes rule:

$$p(\mathbf{r}_0|d_{1:k}) = \frac{g(d_k|\mathbf{r}_0, d_{1:k-1}) p(\mathbf{r}_0|d_{1:k-1})}{f(d_k|d_{1:k-1})} \tag{11}$$

where $f(d_k|d_{1:k-1}) = \int g(d_k|\mathbf{r}_0, d_{1:k-1}) p(\mathbf{r}_0|d_{1:k-1}) d\mathbf{r}_0$ is a normalisation constant. The problem in using (11) is that the likelihood function $g(d_k|\mathbf{r}_0, d_{1:k-1})$ is unknown; only $g(d_k|\boldsymbol{\eta}_0)$ of (6) is known. Fortunately, it is possible to derive an analytic expression for $g(d_k|\mathbf{r}_0, d_{1:k-1})$:

$$g(d_k|\mathbf{r}_0, d_{1:k-1}) = \frac{\vartheta_k^{\kappa_k}\ \Gamma(\kappa_k)}{\vartheta_{k-1}^{\kappa_{k-1}}\ \Gamma(\kappa_{k-1})} \prod_{i=1}^{N} \frac{\rho(\mathbf{r}_0, \mathbf{r}_k^i)^{z_k^i}}{z_k^i!} \tag{12}$$

The Rao-Blackwellised particle filter (RBPF) fully describes the posterior $p_i(\boldsymbol{\eta}_0|d_{1:k})$ by a particle system

$$\mathcal{S}_k^i \equiv \left\{ w_k^{m,i}, \mathbf{r}_{0,k}^{m,i}, \kappa_k^i, \vartheta_k^{m,i} \right\}_{1 \le m \le M}.$$

Here, M is the number of particles, $w_k^{m,i}$ is a (normalised) weight associated with the source position sample $\mathbf{r}_{0,k}^{m,i}$, while κ_k^i and $\vartheta_k^{m,i}$ are the parameters of the corresponding Gamma distribution for the source strength. Initially, at time $k = 0$, the weights are uniform (and equal to $1/M$), $\left\{ \mathbf{r}_{k,0}^{m,i} \right\}$ are the points on a regular grid covering a specified search area, while $\kappa_0^i = \kappa_0$ and $\vartheta_0^{m,i} = \vartheta_0$. The sequential computation of the posterior $p_i(\boldsymbol{\eta}_0|d_{1:k})$ using the RBPF is carried out by a recursive update of the particle system $\mathcal{S}_k^i$ over time.

4. Decentralised formation control

In decentralised multi-robot search, each platform autonomously makes a decision at time t_{k-1} about its next control vector $\mathbf{u}_k^i$, as described in Section 4.1. However, in order to maintain the geometric shape of the formation and thus avoid its break-up, there is a need to impose a form of coordination between the platforms. This will be explained in Section 4.2.

4.1 Selection of individual control vectors

A robot platform i autonomously decides on the control vector $\mathbf{u}_k^i$ using the infotaxis strategy [13], which can be formulated as a partially observed Markov decision process (POMDP) [29]. The elements of POMDP are (i) the information state, (ii) the set of admissible actions and (iii) the reward function. The information state at time t_{k-1} is the posterior density $p_i(\boldsymbol{\eta}_0|d_{1:k-1})$; it accurately specifies the ith platform current knowledge about the source position and its release rate. Admissible actions can be formed with one or multiple steps ahead. A decision in the context of search is the selection of a motion control vector $\mathbf{u}_k^i \in \mathcal{U}$ which will maximise the reward function. According to Section 2, the space of admissible actions $\mathcal{U}$ is continuous with dimensions: linear velocity V, angular velocity Ω and duration of motion T. In order to reduce the computational complexity of numerical optimisation, $\mathcal{U}$ is adopted as a discrete set with only myopic (one step ahead) controls. In addition, $\mathcal{U}$ is time-invariant and identical for all platforms. If $\mathbb{V}$, $\mathbb{O}$ and $\mathbb{T}$ denote the sets of possible discrete-values of V, Ω and T, respectively, then $\mathcal{U}$ is the Cartesian product $\mathbb{V} \times \mathbb{O} \times \mathbb{T}$. The myopic selection of the control vector at time t_k on platform i is expressed as:

$$\mathbf{u}_k^i = \arg\max_{\mathbf{v}\in\mathcal{U}} \mathbb{E}\{\mathcal{D}[p_i(\eta_0|d_{1:k-1}), z_k^i(\mathbf{v})]\} \tag{13}$$

where $\mathcal{D}$ is the reward function and z_k^i is the future concentration measurement collected by the ith platform if the platform moved under the control $\mathbf{v} \in \mathcal{U}$ to position (x_k^i, y_k^i). In reality, this future measurement is not available (the decision has to be made at time t_{k-1}), and therefore the expectation operator $\mathbb{E}$ with respect to the prior measurement PDF features in (13).

Previous studies of search strategies [3, 20] found that the reward function defined as the *reduction of entropy*, results in the most efficient search. Hence, we adopt the expected reward defined as

$$\mathcal{R}_i = \mathbb{E}\{\mathcal{D}[p_i(\boldsymbol{\eta}_0|d_{1:k-1}), z_k^i(\mathbf{v})]\} = H_{k-1}^i - \mathbb{E}\{H_k^i(z_k^i(\mathbf{v}))\} \tag{14}$$

where H_{k-1} is the current differential entropy, defined as

$$H_{k-1}^i = -\int p_i(\eta_0|d_{1:k-1}) \ \ln p_i(\eta_0|d_{1:k-1}) d\boldsymbol{\eta}_0, \tag{15}$$

while $H_k^i(z_k^i(\mathbf{v})) \leq H_{k-1}$ is the future differential entropy (after a hypothetical control vector $\mathbf{v}$ has been applied to collect z_k^i):

$$H_k^i(z_k^i(\mathbf{v})) = -\int p_i(\boldsymbol{\eta}_0|d_{1:k-1}, d_k^i(\mathbf{v})) \ln p_i(\boldsymbol{\eta}_0|d_{1:k-1}, d_k^i(\mathbf{v})) d\boldsymbol{\eta}_0, \tag{16}$$

where $d_k^i = (x_k^i, y_k^i, z_k^i)$. The expectation operator $\mathbb{E}$ in (14) is with respect to the probability mass function $P\{z_k^i|d_{1:k-1}\} = \int \ell(z_k^i|\boldsymbol{\eta}_0) p_i(\boldsymbol{\eta}_0|d_{1:k-1}) d\boldsymbol{\eta}_0$, that is:

$$\mathbb{E}\{H_k^i(z_k^i(\mathbf{v}))\} = \sum_{z_k^i} P\{z_k^i|d_{1:k-1}\} \cdot H_k(z_k^i(\mathbf{v})). \tag{17}$$

Given that $p_i(\boldsymbol{\eta}_0|d_{1:k-1})$ is approximated by a particle system $\mathcal{S}_{k-1}^i$, one can approximately compute H_{k-1}^i, which features in (14), as

$$H^i_{k-1} \approx -\sum_{m=1}^{M} w^{m,i}_{k-1} \ln w^{m,i}_{k-1}. \tag{18}$$

In order to compute $\mathbb{E}\{H^i_k(z^i_k(\mathbf{v}))\}$ of (17), first note that $P\{z^i_k|d_{1:k-1}\} = \mathcal{P}\left(z^i_k; \hat{\mu}^i_{k-1}\right)$, where $\hat{\mu}^i_{k-1}$ is the predicted mean rate of chemical particle encounters at location $\mathbf{r}^i_k$ (where the platform i would move after applying a hypothetical control $\mathbf{v}$), computed based on $d_{1:k-1}$. According to Section 2,

$$\hat{\mu}^i_{k-1} \approx \sum_{m=1}^{M} w^{m,i}_{k-1} \kappa^i_{k-1} \vartheta^{m,i}_{k-1} \rho\left(\mathbf{r}^{m,i}_{0,k-1}, \mathbf{r}^i_k\right) \tag{19}$$

where the product $\kappa^i_{k-1}\vartheta^{m,i}_{k-1}$ approximates the source release rate as the mean of the Gamma distribution with parameters $\left(\kappa^i_{k-1}, \vartheta^{m,i}_{k-1}\right)$. Next, we find the value of $z_{\max}$ as the minimum value of z' such that the cumulative distribution function $\sum_{z=0}^{z'} \mathcal{P}\left(z; \hat{\mu}^i_{k-1}\right)$ is greater than a certain threshold $1-\varepsilon$, where $\varepsilon \ll 1$. The summation (17) is then computed only for $z^i_k = 0, 1, \cdots, z_{\max}$. Computation of $H_k\left(z^i_k(\mathbf{v})\right)$ is carried out according to (18), except that $w^{m,i}_{k-1}$ is replaced with $w^{m,i}_k = w^{m,i}_{k-1} \cdot \mathcal{P}\left(z^i_k; \mu^{m,i}_{k-1}\right)$, where

$$\mu^{m,i}_{k-1} = \kappa^i_{k-1} \vartheta^{m,i}_{k-1} \rho\left(\mathbf{r}^{m,i}_{0,k-1}, \mathbf{r}^i_k\right).$$

Thus, (17) is approximated with

$$\mathbb{E}\left\{H^i_k\left(z^i_k(\mathbf{v})\right)\right\} \approx \sum_{z=0}^{z_{\max}} \mathcal{P}\left(z; \hat{\mu}^i_{k-1}\right) \left[-\sum_{m=1}^{M} w^{m,i}_k \ln w^{m,i}_k\right] \tag{20}$$

Pseudo-code of the routine for the computation of control vector on platform i is given by Algorithm 1.

Algorithm 1 Computation of $\mathbf{u}^i_k$

1: **Input:** $\mathcal{S}^i_{k-1} \equiv \left\{w^{m,i}_{k-1}, \mathbf{r}^{m,i}_{0,k-1}, \kappa^i_{k-1}, \vartheta^{m,i}_{k-1}\right\}_{1 \le m \le M}$,
2: Compute H_{k-1} using (18)
3: Create admissible set $\mathcal{U} = \mathbb{V} \times \mathbb{O} \times \mathbb{T}$
4: **for** every $\mathbf{v} \in \mathcal{U}$ **do**
5: Compute the future platform location $\mathbf{r}^i_k(\mathbf{v})$
6: Compute $\hat{\mu}^i_{k-1}$ using (19)
7: Determine $z_{\max}$ s.t. $\sum_{z=0}^{z_{\max}} \mathcal{P}\left(z; \hat{\mu}^i_{k-1}\right) > 1 - \varepsilon$
8: Compute $\mathbb{E}\left\{H^i_k\left(z^i_k(\mathbf{v})\right)\right\}$ using (20)
9: Calculate the expected reward $\mathcal{R}_i$ using (14)
10: **end for**
11: Find $\mathbf{u}^i_k$ using (13)
12: **Output:** $\mathbf{u}^i_k$

4.2 Cooperative control through consensus

So far, we have explained how platform i would independent of the other platforms in the formation determine the best action for itself, i.e., $\mathbf{u}^i_k$. In general, individual platforms will disagree on the best action, and in the extreme

$\mathbf{u}_1 \neq \mathbf{u}_2 \neq \cdots, \neq \mathbf{u}_N$. In order to maintain the shape of the formation during the motion period (from time t_{k-1} to t_k), the platforms need to reach an agreement on the common action $\mathbf{u}_k$, to be applied to all platforms at the same time. But this is not sufficient; according to the motion model in Section 2, the platforms also need to agree on the formation centroid coordinates and the common heading angle ϕ_{k-1} to be applied in (4).

We apply decentralised cooperative control based on the average consensus [30, 31]. In a network of collaborating agents, consensus is an iterative protocol designed to reach an agreement regarding a certain quantity of interest. Suppose that every platform, as a node in the communication network, initially has an individual scalar value. The goal of average consensus is for every node in the network to compute the average of initial scalar values, in a completely decentralised manner: by communicating only with the neighbours in the communication graph (without knowing the topology of the communication graph).

In the problem we consider, there is not only a single individual scalar value, but six of them. They include three motion control parameters, i.e., for platform i, V_k^i, Ω_k^i and T_k^i, two formation centroid coordinates, i.e., $\overline{x}_{k-1}^i, \overline{y}_{k-1}^i$ and the heading angle of each platform ϕ_{k-1}^i.

Let us denote the scalar value of interest by b_i, that is

$$b_i \in \left\{V_k^i, \Omega_k^i, T_k^i, \overline{x}_{k-1}^i, \overline{y}_{k-1}^i, \phi_{k-1}^i\right\}.$$

Ideally, we want every platform in the formation to compute the mean value $\overline{b} = \frac{1}{N}\sum_{i=1}^N b_i$. If all platforms in the formation were to use identical average values for motion control, centroid coordinates and heading, then their motion would be coordinated (except for process noise, which will be taken care of through vector $\mathbf{B}_{k-1}^i$ in (4)) and the shape of the formation would be maintained (provided $R_{\max}$ is adequate).

Average consensus is an iterative algorithm. At iteration $s = 0$, the node in the communication graph (the robot platform) will initialise its state $b_i(0)$ using either a component of vector $\mathbf{u}_k^i$ (if b_i is a motion control parameter) or the platform pose $\boldsymbol{\theta}_{k-1}^i$ (if b_i is a formation centroid coordinate or heading angle). This value is locally available. The initial values of centroid coordinates are the actual ith platform coordinates, i.e., $\overline{x}_{k-1}^i(0) = x_{k-1}^i$ and $\overline{y}_{k-1}^i(0) = y_{k-1}^i$. At each following iteration $s = 1, 2, \cdots$, each platform updates its state with a linear combination of its own state and the states of its current neighbours. Let us denote the set of current neighbours of platform i by $\mathcal{J}_i$. Then [30]:

$$b_i(s) = \left(1 - \frac{|\mathcal{J}_i|}{N}\right) b_i(s-1) + \frac{1}{N}\sum_{j\in\mathcal{J}_i} b_j(s-1) \tag{21}$$

where $|\mathcal{J}_i|$ is the number of neighbours of platform i. This particular linear combination is based on the so-called *maximum degree weights* [32]. Other weights can be also used. It can be shown that if the communication graph is connected, the values $b_i(s)$ after many iterations converge to the mean $\overline{b}$ [32].

The search continues until the global stopping criterion is satisfied. The local stopping criterion is calculated on each platform independently based on the spread of the local positional particles $\left\{\mathbf{r}_{0,k}^{m,i}\right\}$, measured by the square-root of the trace of its sample covariance matrix $\mathbf{C}_k$. For example, if the spread of particles on platform i is smaller than a certain threshold ϖ, then the local stopping criterion is satisfied and is given a value of one, otherwise it is zero. This local stopping criterion value (zero or one) becomes the initial state of the global stopping criterion on platform i, denoted $\sigma_i(0)$:

$$\sigma_i(0) = \begin{cases} 1 & \text{if} \sqrt{tr[\mathbf{C}_k]} < \varpi \\ 0 & \text{otherwise} \end{cases} \tag{22}$$

The global stopping criterion is computed on each platform using the average consensus algorithm, using (21), but with b_i replaced by σ_i. After a sufficient number of iterations, S, platform i decides to stop the search if at least one of the platforms in the formation has reached the local stopping criterion, that is, if $\sigma_i(S) > 0$.

We point out that both estimation and control are based on the consensus algorithm. While the cooperative control is using the *average* consensus (21), the decentralised measurement dissemination of Section 3 achieves the consensus on the set of measurements at time k. The consensus algorithm is iterative, and hence its convergence properties are very important. First note that, although the network topology changes with time (as the robots move while searching for the source), during the short interval of time when the exchange of information takes place, the topology can be considered as *time-invariant*. Furthermore, assuming bidirectional communication between the robots in formation, the network topology can be represented by an undirected graph. The convergence of the consensus algorithm for a time-invariant undirected communication topology is guaranteed if the graph is connected [31–33]. Note that this theoretical result is valid for an infinite number of iterations. In practice, if the communication graph at some point of time is not connected, or if an insufficient number of consensus iterations are performed, it may happen that one or more robots are lost (they could re-join the formation only by coincidence). This event, however, does not mean that the search mission has failed: the emitting source will be found eventually, albeit by a smaller formation in possibly longer interval of time.

5. Numerical results

The proposed search algorithm has been applied to an experimental dataset, collected by COANDA Research & Development Corporation using their large recirculating water channel. The emitting source was releasing fluorescent dye at a constant rate from a narrow tube. The dataset comprises a sequence of 340 frames of instantaneous concentration field measurements in the vertical plane and is sampled at every 10/23 s. The size of a frame is 49×49 pixels, where a pixel corresponds to a square area of $2.935 \times 2.935\,\text{mm}^2$. As the size of the data is relatively small, we follow the approach used in [24]: upscale each frame by a factor of 3 using bicubic interpolation and place the result in the top left corner of a 500×500 search area. A measurement obtained by a platform is, thus, the integer value of the concentration of the dye taken from the closest spatial and temporal sample from the experimental data.

An example of the search algorithm running on the experimental data is shown in **Figure 2**. All physical quantities are in arbitrary units (a.u.). The following environmental/sensing parameters were used: $D = 1$, $\tau = 250$, $U = 0$, $a = 1$ and $t_0 = 1$. Algorithm parameters are selected as follows: $\kappa_0 = 3$, $\vartheta_0 = 5.2$, number of particles $M = 25^2$, $\mathbb{V} = \{1\}$, $\mathbb{O} = \{-3, -2, -1, 0, 1, 2, 3\}$ degrees per unit of time and $\mathbb{T} = \{0.5, 1, 2, 4, 8, 16, 32, 64\}$. The number of iterations, both for the exchange of measurement triples and in the consensus algorithm, was fixed to 30. The local search stopping threshold was $\varpi = 3$.

Figure 2 displays the top-down view of the search progress at step indices $k = 0, 12, 22, 32$. The formation consists of $N = 7$ platforms, whose trajectories are shown in different colours. The search algorithm terminated at $K = 33$. Note that the plume size is much smaller than the search area. Panels **(a)–(c)** of **Figure 2** show the particles before resampling: the particles are placed on a regular grid, thus mimicking a grid-based approach, with the value of particle weights indicated by the grey-scale intensity plot (white means a zero weight). This provides a good visual representation of the posterior $p(\mathbf{r}_0|d_{1:k})$. Panel **(d)** shows the situation after a non-zero concentration measurement was collected by the search team. The positional particles have been resampled at this point of time and moved closer to the true source location.

Using 200 Monte Carlo simulations, the mean search time for the algorithm was 2525 a.u., with a 5th and 95th quantile of 1840 and 3445 a.u., respectively. Note that in all simulations the formation started from the bottom right hand corner indicated in **Figure 2(a)**.

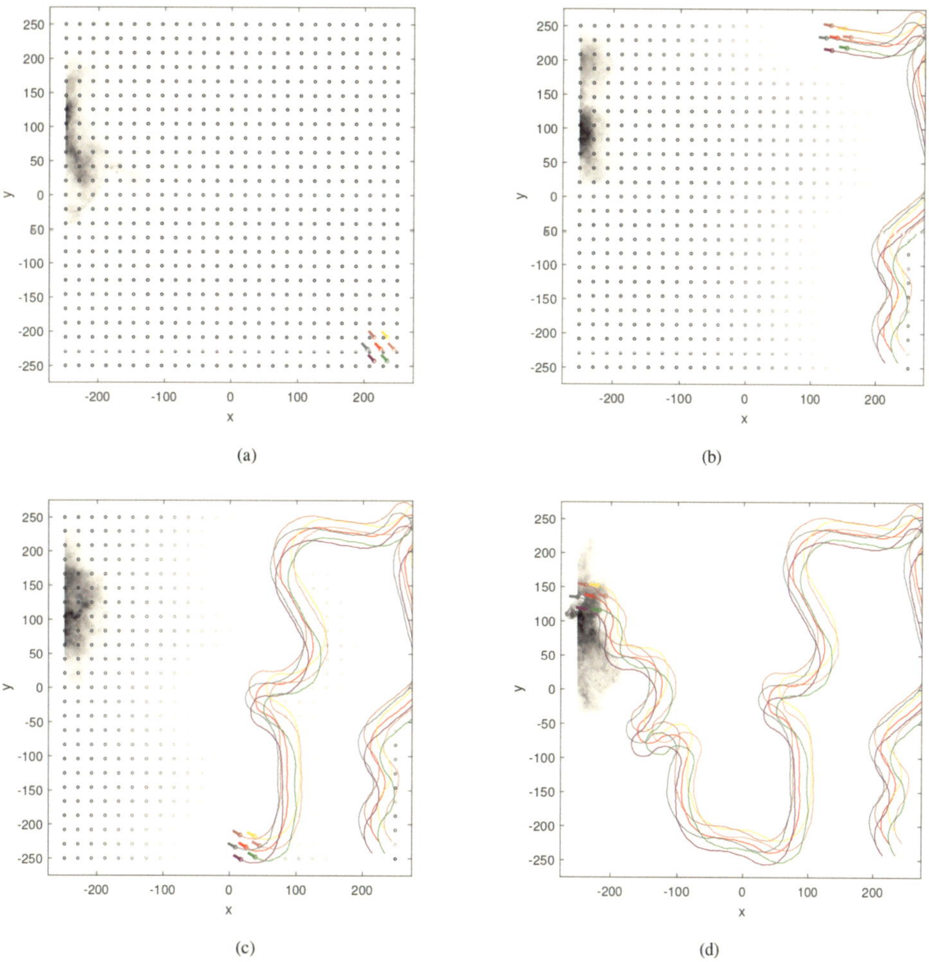

Figure 2.
Experimental dataset: an illustrative run of the decentralised multi-robot search using N = 7 platforms. Graphs (a)–(d) show the positions and trajectories of the platforms at step indices k = 0,12,22 and 32, respectively. The concentration of the plume is represented in grey-scale (darker colours represent higher concentration).

6. Conclusions

The chapter presented a decentralised infotaxic search algorithm for a group of autonomous robotic platforms. The algorithm allows the platforms to search and locate a source of hazardous emissions in a coordinated manner without the need for a centralised fusion and control system. More precisely, this distributed coordination is achieved only by local exchange of measurement data between neighbouring platforms. Similarly, the movement decisions taken by the platforms were reached using a distributed average consensus algorithm over the whole formation. The key aspect is that individual platforms only require knowledge of their neighbours; the global knowledge of the communication network topology is unnecessary. An advantage of adopted distributed framework is that all platforms are treated equally, making the proposed search algorithm scalable and robust to the failure of a single platform. Numerical results using experimental data confirmed the robust performance of the algorithm. The main limitation of the algorithm is that the environmental parameters (such as diffusivity, the average direction and speed of the wind, particle lifetime), must be known. Future work will explore sensitivity to parametrisation and will aim to develop a team of "search and rescue" robots for further experimentation in realistic environments.

Acknowledgements

This research was supported in part by the Defence Science and Technology Group through its Strategic Research Initiative on Trusted Autonomous Systems.

Author details

Branko Ristic* and Christopher Gilliam
School of Engineering, RMIT University, Melbourne, Australia

*Address all correspondence to: branko.ristic@rmit.edu.au

References

[1] Hutchinson M, Liu C, Chen W-H. Information-based search for an atmospheric release using a mobile robot: Algorithm and experiments. IEEE Transactions on Control Systems Technology. 2018;**99**:1-15

[2] Ristic B, Morelande M, Gunatilaka A. Information driven search for point sources of gamma radiation. Signal Processing. 2010;**90**:1225-1239

[3] Ristic B, Skvortsov A, Gunatilaka A. A study of cognitive strategies for an autonomous search. Information Fusion. 2016;**28**:1-9

[4] Haley KB, Stone LD, editors. Search Theory and Applications. Nato Conference Series. New York: Plenum Press; 1980

[5] Halford SE. How do site-specific DNA-binding proteins find their targets? Nucleic Acids Research. 2004; **32**(10):3040-3052

[6] Fauchald P, Tveraa T. Using first-passage time in the analysis of area-restricted search and habitat selection. Ecology. 2003;**84**(2):282-288

[7] Viswanathan GM, Afanasyev V, Buldyrev SV, Murphy EJ, Prince PA, Satnley HE. Levy flight search patterns of wandering albatrosses. Nature. 1996; **381**:413-415

[8] Shlesinger MF. Mathematical physics: Search research. Nature. 2006; **443**(7109):281-282

[9] Bayat B, Crasta N, Crespi A, Pascoal AM, Ijspeert A. Environmental monitoring using autonomous vehicles: A survey of recent searching techniques. Current Opinion in Biotechnology. 2017;**45**:76-84

[10] Dunbabin M, Marques L. Robots for environmental monitoring: Significant advancements and applications. IEEE Robotics and Automation Magazine. 2012;**19**(1):24-39

[11] Ishida H, Wada Y, Matsukura H. Chemical sensing in robotic applications: A review. IEEE Sensors Journal. 2012;**12**(11):3163-3173

[12] Masson J-B, Bailly-Bachet M, Vergassola M. Chasing information to search in random environments. Journal of Physics A: Mathematical and Theoretical. 2009;**42**:434009

[13] Vergassola M, Villermaux E, Shraiman BI. 'Infotaxis' as a strategy for searching without gradients. Nature. 2007;**445**(25):406-409

[14] Barbieri C, Cocco S, Monasson R. On the trajectories and performance of infotaxis, an information-based greedy search algorithm. Europhysics Letters. 2011;**94**(2):20005

[15] Fatès N. Collective infotaxis with reactive amoebae: A note on a simple bio-inspired mechanism. In: International Conference on Cellular Automata; 2016. pp. 157-165

[16] Hajieghrary H, Ani Hsieh M, Schwartz IB. Multi-agent search for source localization in a turbulent medium. Physics Letters A. 2016; **380**(20):1698-1705

[17] Hajieghrary H, Tomas AF, Hsieh MA. An information theoretic source seeking strategy for plume tracking in 3D turbulent fields. In: Proceedings of the IEEE International Symposium on Safety, Security, and Rescue Robotics (SSRR); 2015. pp. 1-8

[18] Hein AM, McKinley SA. Sensing and decision-making in random search. Proceedings of the National Academy of Sciences of the United States of America. 2012;**109**:12070-12074

[19] Hutchinson M, Oh H, Chen W-H. Entrotaxis as a strategy for autonomous search and source reconstruction in turbulent conditions. Information Fusion. 2018;**42**:179-189

[20] Masson J-B. Olfactory searches with limited space perception. Proceedings of the National Academy of Sciences of the United States of America. 2013;**110**(28): 11261-11266

[21] Moraud EM, Martinez D. Effectiveness and robustness of robot infotaxis for searching in dilute conditions. Frontiers in Neurorobotics. 2010;**4**:1-8

[22] Ristic B, Skvortsov A, Walker A. Autonomous search for a diffusive source in an unknown structured environment. Entropy. 2014;**16**(2): 789-813

[23] Voges N, Chaffiol A, Lucas P, Martinez D. Reactive searching and infotaxis in odor source localization. PLoS Computational Biology. 2014;**10**(10):e1003861

[24] Ristic B, Angley D, Moran B, Palmer JL. Autonomous multi-robot search for a hazardous source in a turbulent environment. Sensors. 2017;**17**(4):918

[25] Hlinka O, Hlawatsch F, Djuric PM. Distributed particle filtering in agent networks: A survey, classification, and comparison. IEEE Signal Processing Magazine. 2013;**30**(1):61-81

[26] Doucet A, De Freitas N, Murphy K, Russell S. Rao-Blackwellised particle filtering for dynamic Bayesian networks. In: Proceedings of the Sixteenth Conference on Uncertainty in Artificial Intelligence; 2000. pp. 176-183

[27] Ristic B, Gunatilaka A, Wang Y. Rao–Blackwell dimension reduction applied to hazardous source parameter estimation. Signal Processing. 2017;**132**: 177-182

[28] Gelman A, Carlin JB, Stern HS, Rubin DB. Bayesian Data Analysis. 2nd ed. Boca Raton FL: CRC Press; 2004

[29] Chong EKP, Kreucher C, Hero AO. Chapter 8. POMDP approximation using simulation and heuristics. In: Hero AO, Castanon D, Cochran D, Kastella K, editors. Foundations and Applications of Sensor Management. New York: Springer; 2008

[30] Xiao L, Boyd S. Fast linear iterations for distributed averaging. Systems & Control Letters. 2004;**53**(1):65-78

[31] Ren W, Beard RW, Atkins EM. Information consensus in multivehicle cooperative control. IEEE Control Systems. 2007;**27**(2):71-82

[32] Xiao L, Boyd S, Lall S. A scheme for robust distributed sensor fusion based on average consensus. In: Proceedings of the 4th International Symposium on Information Processing in Sensor Networks; 2005. p. 9

[33] Olfati-Saber R, Fax JA, Murray RM. Consensus and cooperation in networked multi-agent systems. Proceedings of the IEEE. 2007;**95**(1): 215-233

Chapter 3

Vision-Based Autonomous Control Schemes for Quadrotor Unmanned Aerial Vehicle

Archit Krishna Kamath, Vibhu Kumar Tripathi and Laxmidhar Behera

Abstract

This chapter deals with the development of vision-based sliding mode control strategies for a quadrotor system that would enable it to perform autonomous tasks such as take-off, landing and visual inspection of structures. The aim of this work is to provide a basic understanding of the quadrotor dynamical model, key concepts in image processing and a detailed description of the sliding mode control, a widely used robust non-linear control scheme. Extensive MATLAB simulations are presented to enhance the understanding of the controller on the quadrotor system subjected to bounded disturbances and uncertainties. The vision algorithms developed in this chapter would provide the necessary reference trajectory to the controller enabling it to exercise control over the system. This work also describes, in brief, the implementation of the developed control and vision algorithms on the DJI Matrice 100 to present real-time experimental data to the readers of this chapter.

Keywords: quadrotor dynamical control, unmanned aerial vehicle, vision-based control, sliding mode control

1. Introduction

In the past few years, the interest in unmanned aerial vehicle (UAV) has been growing strongly. The possibility of removing human pilots from danger as well as the size and cost of UAVs are indeed very attractive but have to be compared to the performances attained by human-piloted vehicles in terms of mission capabilities, efficiency and flexibility. The design of flight controllers able to offer to UAVs an accurate and robust control is an important step in the design of fully autonomous vehicles. In practical operations, fixed-wing UAVs have been used for years in routine surveillance missions but their lack of stationary flight capability has shifted the focus to vertical take-off and landing (VTOL) vehicles offering the possibility of being launched from virtually anywhere along with the ability to hover above a target. Several designs are available when it comes to VTOL vehicles; however, the quadrotor configuration presented in this chapter offers all the advantages of VTOL vehicles along with an increased payload capacity, a stability in hover inherent to its design (while it is the hardest flight condition to maintain for conventional helicopter) as well as an increased maneuverability [1].

In this work, the vision-based position and altitude tracking control of a quadrotor UAV is considered. This would be then on used to align the drone to the center of a pre-defined landing pad marker on which the quadrotor would autonomously land. In practical missions, the stability of the quadrotor is easily affected by abrupt changes in the input commands. The flight controller that is designed must be capable in offering an accurate and robust control to the quadrotor. The controller demonstrated in this chapter is the sliding mode controller (SMC). The sliding mode control (SMC) technique, being a non-linear control technique, has found great applications in offering robust control solutions for handling quadrotors [2–7]. This chapter will briefly describe the process of implementing a vision algorithm alongside a classical SMC for autonomous landing of the quadrotor on a stationary platform.

2. Quadrotor configuration

The quadrotor UAV is a highly non-linear, 6 DoF, Multi-Input-Multi-Output (MIMO) and under-actuated system [8]. One can describe the vehicle as having four propellers in cross configuration as shown in **Figure 1**. Quadrotor motion is controlled by varying the speed of the four rotors. A quadrotor has two sets of clockwise and two sets of counter-clockwise rotating propellers to neutralize the effective aerodynamic drag. Vertical movement of the quadrotor system is controlled by simultaneously increasing or decreasing the thrust of all rotors. Yawing motion is created by proportionally varying the speeds of counter-clockwise rotating propellers and the rolling and pitching motions are created by applying differential thrust forces on opposite rotors of the quadrotor [9].

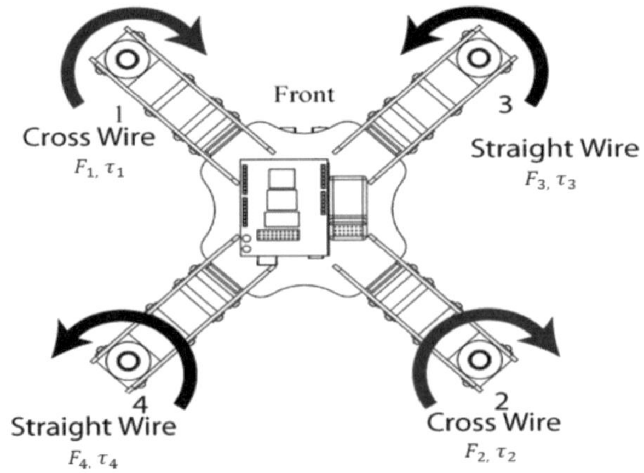

Figure 1.
Quadrotor UAV configuration.

3. Quadrotor mathematical model

The quadrotor dynamics, also called the equations of motion, are a set of 6 s order differential equations. The quadrotor being a 6 DoF plant, a total of 12 states are required to describe its motion completely. These 12 states are described using the 6 equations of motion. These play a vital role in controller design and would be extensively used in the subsequent sections of this chapter.

The kinematic and dynamic models of a quadrotor will be derived based on a Newton-Euler formalism with the following assumptions [10]:

- The quadrotor structure is assumed to be rigid and symmetrical.
- The center of gravity of the quadrotor coincides with the body fixed frame origin.
- The propellers are rigid.
- Thrust and drag are proportional to the square of propeller's speed.

The first step in developing the quadrotor kinematic model is to describe the different frames of references associated with the system. It is necessary to use these coordinate systems for the following reasons:

1. Newton's equations of motion are given the coordinate frame attached to the quadrotor.
2. Aerodynamics forces and torques are applied in the body frame.
3. On-board sensors like accelerometers and rate gyros measure information with respect to the body frame. Alternatively, GPS measures position, ground speed, and course angle with respect to the inertial frame.
4. Most mission requirements like loiter points and flying trajectories are specified in the inertial frame. In addition, map information is also given in an inertial frame.

In this case, we describe a total of frames, namely: inertial frame (F^i), the vehicle frame (F^v), the vehicle frame-1 (F^{v1}), the vehicle frame-2 (F^{v2}), and the body frame (F^b). The inertial frame is fixed at a point at ground level and uses the N-E-D notation, where N points towards north direction, E points towards east direction and D points towards earth. On the other hand, the body frame is at the center of quadrotor body, with its x axis pointing towards the front of the quadrotor, y axis pointing towards the left of the quadrotor and the z axis pointing towards the ground. The vehicle frame has an axis parallel to the inertial frame but has the origin shifted to the quadrotor's center of gravity. Vehicle frame's yaw is adjusted to match the quadrotor's yaw to get the vehicle frame-1 frame which is then pitch adjusted to get the vehicle frame-2. Finally the body frame is obtained by adjusting the roll of the vehicle frame-2.

The transformation from inertial to vehicle frame is just a simple translation. On the other hand, the transformation from vehicle to body frame is given by a rotation matrix $R_v^b(\phi,\theta,\psi)$, given by:

$$\left.\begin{aligned} R_v^b(\phi,\theta,\psi) &= R_{v2}^b(\phi)R_{v1}^{v2}(\theta)R_v^{v1}(\psi) = \\ &\begin{bmatrix} 1 & 0 & 0 \\ 0 & \cos(\phi) & \sin(\phi) \\ 0 & -\sin(\phi) & \cos(\phi) \end{bmatrix}\begin{bmatrix} \cos(\theta) & 0 & -\sin(\theta) \\ 0 & 1 & 0 \\ \sin(\theta) & 0 & \cos(\theta) \end{bmatrix}\begin{bmatrix} \cos(\psi) & \sin(\psi) & 0 \\ -\sin(\psi) & \cos(\psi) & 0 \\ 0 & 0 & 1 \end{bmatrix} \\ &= \begin{bmatrix} c_\theta c_\psi & s_\phi s_\theta c_\psi - c_\phi c_\psi & c_\phi s_\theta c_\psi + s_\phi s_\psi \\ c_\theta s_\psi & s_\phi s_\theta s_\psi + c_\phi c_\psi & c_\phi s_\theta s_\psi - s_\phi c_\psi \\ -s_\theta & s_\phi c_\theta & c_\phi c_\theta \end{bmatrix} \end{aligned}\right\} \tag{1}$$

where, ϕ, θ and ψ represent the roll, pitch and yaw angles of the quadrotor measured in the vehicle frame-2, vehicle frame-1 and vehicle frame respectively. In addition to these Euler angles, the quadrotor is associated with several other state variables that describe its position, linear velocity and angular velocities. These are described as:

1. x—The inertial (north) position of the quadrotor.

2. y—The inertial (east) position of the quadrotor.

3. z—The altitude of the aircraft.

4. u—The body frame velocity in x direction in body frame.

5. v—The body frame velocity in y direction in body frame.

6. w—The body frame velocity in z direction in body frame.

7. p—The roll rate measured in body frame.

8. q—The pitch rate measured in body frame.

9. r—The yaw rate measured in body frame.

Hence a total of 12 states are used to describe the motion of the quadrotor in the 3D space.

3.1 Kinematic model

The position derivatives $(\dot{x}, \dot{y}, \dot{z})$ are inertial frame quantities and velocities (u, v, w) are in the body frame. They can be related through the transformation matrix as follows [11]:

$$\left.\begin{aligned} \frac{d}{dt}\begin{bmatrix} x \\ y \\ z \end{bmatrix} &= R_b^v \begin{bmatrix} u \\ v \\ w \end{bmatrix} \\ &= \left(R_v^b\right)^T \begin{bmatrix} u \\ v \\ w \end{bmatrix} \\ &= \begin{bmatrix} c_\theta c_\psi & s_\phi s_\theta c_\psi - c_\phi c_\psi c_\phi s_\theta c_\psi + s_\phi s_\psi \\ c_\theta s_\psi & s_\phi s_\theta s_\psi + c_\phi c_\psi c_\phi s_\theta s_\psi - s_\phi c_\psi \\ -s_\theta & s_\phi c_\theta c_\phi c_\theta \end{bmatrix} \begin{bmatrix} u \\ v \\ w \end{bmatrix} \end{aligned}\right\} \quad (2)$$

The relationship between absolute angles ϕ, θ, and ψ, and the angular rates p, q, and r is also complicated by the fact that these quantities are defined in different coordinate frames. The angular rates are defined in the body frame F^b, whereas the roll angle ϕ is defined in F^{v2}, the pitch angle θ is defined in F^{v1}, and the yaw angle ψ is defined in the vehicle frame F^v.

We need to relate p, q, and r to $\dot{\phi}$, $\dot{\theta}$, and $\dot{\psi}$. Since $\dot{\phi}$, $\dot{\theta}$ and $\dot{\psi}$ are small and noting that $R_{v1}^{v}(\dot{\psi})$, $R_{v2}^{v1}(\dot{\theta})$ and $R_{b}^{v2}(\dot{\phi})$ are all identity matrices, we get:

$$\left.\begin{aligned} \begin{bmatrix} p \\ q \\ r \end{bmatrix} &= R_b^{v2}(\dot{\phi})\begin{bmatrix} \dot{\phi} \\ 0 \\ 0 \end{bmatrix} + R_b^{v2}(\dot{\phi})R_{v2}^{v1}(\dot{\theta})\begin{bmatrix} 0 \\ \dot{\theta} \\ 0 \end{bmatrix} + R_b^{v2}(\dot{\phi})R_{v2}^{v1}(\dot{\theta})R_{v1}^{v}(\dot{\psi})\begin{bmatrix} 0 \\ 0 \\ \dot{\psi} \end{bmatrix} \\ &= \begin{bmatrix} 1 & 0 - \sin(\theta) & \\ 0 & \cos(\phi)\ \sin(\phi)\cos(\theta) & \\ 0 & -\sin(\phi)\ \cos(\phi)\cos(\theta) & \end{bmatrix}\begin{bmatrix} \dot{\phi} \\ \dot{\theta} \\ \dot{\psi} \end{bmatrix} \end{aligned}\right\} \tag{3}$$

Inverting this, we get:

$$\begin{bmatrix} \dot{\phi} \\ \dot{\theta} \\ \dot{\psi} \end{bmatrix} = \begin{bmatrix} 1 & \sin(\phi)\tan(\theta) & \cos(\phi)\tan(\theta) \\ 0 & \cos(\phi) & -\sin(\phi) \\ 0 & \sin(\phi)\sec(\theta) & \cos(\phi)\sec(\theta) \end{bmatrix}\begin{bmatrix} p \\ q \\ r \end{bmatrix} \tag{4}$$

3.2 Dynamic model

Let $\mathbf{v}$ be the velocity vector of the quadrotor. Newton's laws of motion hold good for inertial frames of references only. On applying these to a transnational frame, the equation modifies as follows:

$$m\frac{d\mathbf{v}}{dt_i} = f \tag{5}$$

where m is the mass of the quadrotor, f is the total applied to the quadrotor, and $\frac{d}{dt_i}$ is the time derivative in the inertial frame. From the equation of Coriolis, we have:

$$m\frac{d\mathbf{v}}{dt_i} = m\left(\frac{d\mathbf{v}}{dt_b} + \omega_{b/i} \times \mathbf{v}\right) = f \tag{6}$$

where $\omega_{b/i}$ is the angular velocity of the air-frame with respect to the inertial frame. Since the control force is computed and applied in the body coordinate system, and since ω is measured in body coordinates, we will express the above equation in body coordinates, where $\mathbf{v}^b = (u, v, w)^T$, and $\omega_{b/i}^b = (p, q, r)^T$. Therefore, in body coordinates the above equation becomes:

$$\begin{bmatrix} \dot{u} \\ \dot{v} \\ \dot{w} \end{bmatrix} = \begin{bmatrix} rv - qw \\ pw - ru \\ qu - pv \end{bmatrix} + \frac{1}{m}\begin{bmatrix} f_x \\ f_y \\ f_z \end{bmatrix} \tag{7}$$

where $\left[f_x f_y f_z\right]^T = f$.

For rotational motion, Newton's second law states that:

$$\frac{d\mathbf{h}^b}{dt_i} = \mathbf{m} \tag{8}$$

where $\mathbf{h}$ is the angular momentum and $\mathbf{m}$ is the applied torque. Using the equation of Coriolis we have:

$$\frac{d\mathbf{h}}{dt_i} = \left(\frac{d\mathbf{h}}{dt_b} + \omega_{b/i} \times \mathbf{h}\right) = \mathbf{m} \tag{9}$$

Again, the above equation is most easily resolved in body coordinates where $\mathbf{h}^b = \mathbf{J}\omega_{b/i}^b$, where $\mathbf{J}$ is the constant inertia matrix given by:

$$\mathbf{J} = \begin{bmatrix} J_x & -J_{xy} & -J_{xz} \\ -J_{xy} & J_y & -J_{yz} \\ -J_{xz} & -J_{yz} & J_z \end{bmatrix} \tag{10}$$

As we use a quadrotor with a symmetric frame about all three axes, $J_{xy} = J_{yz} = J_{xz} = 0$. Hence, $\mathbf{J}$ becomes:

$$\mathbf{J} = \begin{bmatrix} J_x & 0 & 0 \\ 0 & J_y & 0 \\ 0 & 0 & J_z \end{bmatrix} \tag{11}$$

Defining $\mathbf{m}^b = \left[\tau_\phi \tau_\theta \tau_\psi\right]^T$ we can write Eq. (9) in the body coordinates as:

$$\mathbf{J}\begin{bmatrix} \dot{p} \\ \dot{q} \\ \dot{r} \end{bmatrix} = \begin{bmatrix} 0 & r & -q \\ -r & 0 & p \\ p & -q & 0 \end{bmatrix}\begin{bmatrix} J_x & 0 & 0 \\ 0 & J_y & 0 \\ 0 & 0 & J_z \end{bmatrix}\begin{bmatrix} p \\ q \\ r \end{bmatrix} + \begin{bmatrix} \tau_\phi \\ \tau_\theta \\ \tau_\psi \end{bmatrix} \tag{12}$$

Hence:

$$\begin{bmatrix} \dot{p} \\ \dot{q} \\ \dot{r} \end{bmatrix} = \begin{bmatrix} \frac{J_y - J_z}{J_x} qr \\ \frac{J_z - J_x}{J_y} pr \\ \frac{J_x - J_y}{J_z} qp \end{bmatrix} + \begin{bmatrix} \frac{1}{J_x}\tau_\phi \\ \frac{1}{J_y}\tau_\theta \\ \frac{1}{J_z}\tau_\psi \end{bmatrix} \tag{13}$$

To summarize, from Eqs. (2)–(13), we obtain the 6 DoF equation of a quadrotor and is given as follows:

$$\left.\begin{aligned}
\begin{bmatrix} \dot{x} \\ \dot{y} \\ \dot{z} \end{bmatrix} &= \begin{bmatrix} c_\theta c_\psi & s_\phi s_\theta c_\psi - c_\phi c_\psi & c_\phi s_\theta c_\psi + s_\phi s_\psi \\ c_\theta s_\psi & s_\phi s_\theta s_\psi + c_\phi c_\psi & c_\phi s_\theta s_\psi - s_\phi c_\psi \\ -s_\theta & s_\phi c_\theta & c_\phi c_\theta \end{bmatrix}\begin{bmatrix} u \\ v \\ w \end{bmatrix} \\
\begin{bmatrix} \dot{\phi} \\ \dot{\theta} \\ \dot{\psi} \end{bmatrix} &= \begin{bmatrix} 1 & \sin(\phi)\tan(\theta) & \cos(\phi)\tan(\theta) \\ 0 & \cos(\phi) & -\sin(\phi) \\ 0 & \sin(\phi)\sec(\theta) & \cos(\phi)\sec(\theta) \end{bmatrix}\begin{bmatrix} p \\ q \\ r \end{bmatrix} \\
\begin{bmatrix} \dot{u} \\ \dot{v} \\ \dot{w} \end{bmatrix} &= \begin{bmatrix} rv - qw \\ pw - ru \\ qu - pv \end{bmatrix} + \frac{1}{m}\begin{bmatrix} f_x \\ f_y \\ f_z \end{bmatrix} \\
\begin{bmatrix} \dot{p} \\ \dot{q} \\ \dot{r} \end{bmatrix} &= \begin{bmatrix} \frac{J_y - J_z}{J_x} qr \\ \frac{J_z - J_x}{J_y} pr \\ \frac{J_x - J_y}{J_z} qp \end{bmatrix} + \begin{bmatrix} \frac{1}{J_x}\tau_\phi \\ \frac{1}{J_y}\tau_\theta \\ \frac{1}{J_z}\tau_\psi \end{bmatrix}
\end{aligned}\right\} \tag{14}$$

3.3 Forces and moments

The objective of this section is to describe the forces and torques that act on the quadrotor. Since there are no aerodynamic lifting surfaces, we will assume that the aerodynamic forces and moments are negligible. The forces and moments are primarily due to gravity and the four propellers.

As seen in **Figure 1**, each motor produces a force F and a torque τ. The total force acting on the quadrotor is given by:

$$F = F_1 + F_2 + F_3 + F_4 \tag{15}$$

The rolling torque is produced by the force difference between the motor pair 1–4 and 2–3 and is given as:

$$\tau_\phi = L(F_1 + F_4) - L(F_2 + F_3) \tag{16}$$

Similarly, the pitching torque is produced by the force difference between the motor pair 1–3 and 2–4 and is given as:

$$\tau_\theta = L(F_1 + F_3) - L(F_2 + F_4) \tag{17}$$

Due to Newton's third law, the drag of the propellers produces a yawing torque on the body of the quadrotor. The direction of the torque will be in the opposite direction of the motion of the propeller. Therefore the total yawing torque is given by:

$$\tau_\psi = \tau_1 + \tau_2 - \tau_3 - \tau_4 \tag{18}$$

The lift and drag produced by the propellers is proportional to the square of the angular velocity. We will assume that the angular velocity is directly proportional to the pulse width modulation command sent to the motor. Therefore, the force and torque of each motor can be expressed as:

$$\left.\begin{aligned} F_* &= K_1\delta_* \\ \tau_* &= K_2\delta_* \end{aligned}\right\} \tag{19}$$

where K_1 and K_2 are constants that are determined experimentally, δ_* is the motor command signal, and *—represents 1, 2, 3, and 4. Therefore, the forces and torques on the quadrotor can be written in matrix form as:

$$\begin{bmatrix} F \\ \tau_\phi \\ \tau_\theta \\ \tau_\psi \end{bmatrix} = \begin{bmatrix} K_1 & K_1 & K_1 & K_1 \\ LK_1 & -LK_1 & -LK_1 & LK_1 \\ LK_1 & -LK_1 & LK_1 & -LK_1 \\ K_2 & K_2 & -K_2 & -K_2 \end{bmatrix} \begin{bmatrix} \delta_1 \\ \delta_2 \\ \delta_3 \\ \delta_4 \end{bmatrix} = M \begin{bmatrix} \delta_1 \\ \delta_2 \\ \delta_3 \\ \delta_4 \end{bmatrix} \tag{20}$$

The control strategies derived in subsequent sections will specify forces and torques. The actual motors commands can be found as:

$$\begin{bmatrix} \delta_1 \\ \delta_2 \\ \delta_3 \\ \delta_4 \end{bmatrix} = M^{-1} \begin{bmatrix} F \\ \tau_\phi \\ \tau_\theta \\ \tau_\psi \end{bmatrix} \tag{21}$$

Note that the pulse width modulation commands are required to be between zero and one.

In addition to the force exerted by the motor, gravity also exerts a force on the quadrotor. In the vehicle frame F^v, the gravity force acting on the center of mass is given by:

$$\mathbf{f}_g^b = \begin{bmatrix} 0 \\ 0 \\ mg \end{bmatrix} \tag{22}$$

Hence, transforming $\mathbf{f}_v^b$, we get:

$$\left.\begin{aligned} \mathbf{f}_v^b &= R_b^v \begin{bmatrix} 0 \\ 0 \\ mg \end{bmatrix} \\ &= \begin{bmatrix} -mg\sin(\theta) \\ mg\cos(\theta)\sin(\phi) \\ mg\cos(\theta)\cos(\phi) \end{bmatrix} \end{aligned}\right\} \tag{23}$$

Therefore, transforming equation set (14), we get:

$$\begin{bmatrix} \dot{x} \\ \dot{y} \\ \dot{z} \end{bmatrix} = \begin{bmatrix} c_\theta c_\psi & s_\phi s_\theta c_\psi - c_\phi c_\psi & c_\phi s_\theta c_\psi + s_\phi s_\psi \\ c_\theta s_\psi & s_\phi s_\theta s_\psi + c_\phi c_\psi & c_\phi s_\theta s_\psi - s_\phi c_\psi \\ -s_\theta & s_\phi c_\theta & c_\phi c_\theta \end{bmatrix} \begin{bmatrix} u \\ v \\ w \end{bmatrix} \tag{24}$$

$$\begin{bmatrix} \dot{\phi} \\ \dot{\theta} \\ \dot{\psi} \end{bmatrix} = \begin{bmatrix} 1 & \sin(\phi)\tan(\theta) & \cos(\phi)\tan(\theta) \\ 0 & \cos(\phi) & -\sin(\phi) \\ 0 & \sin(\phi)\sec(\theta) & \cos(\phi)\sec(\theta) \end{bmatrix} \begin{bmatrix} p \\ q \\ r \end{bmatrix} \tag{25}$$

$$\begin{bmatrix} \dot{u} \\ \dot{v} \\ \dot{w} \end{bmatrix} = \begin{bmatrix} rv - qw \\ pw - ru \\ qu - pv \end{bmatrix} + \begin{bmatrix} -g\sin(\theta) \\ g\cos(\theta)\sin(\phi) \\ g\cos(\theta)\cos(\phi) \end{bmatrix} + \frac{1}{m}\begin{bmatrix} 0 \\ 0 \\ -F \end{bmatrix} \tag{26}$$

$$\begin{bmatrix} \dot{p} \\ \dot{q} \\ \dot{r} \end{bmatrix} = \begin{bmatrix} \frac{J_y - J_z}{J_x} qr \\ \frac{J_z - J_x}{J_y} pr \\ \frac{J_x - J_y}{J_z} qp \end{bmatrix} + \begin{bmatrix} \frac{1}{J_x}\tau_\phi \\ \frac{1}{J_y}\tau_\theta \\ \frac{1}{J_z}\tau_\psi \end{bmatrix} \tag{27}$$

Eqs. (24)–(27) represent the complete non-linear model of the quadrotor. However, they are not appropriate for control design for several reasons. The first reason is that they are too complicated to gain significant insight into the motion of the quadrotor. The second reason is that the position and orientation are relative to the inertial world fixed frame, whereas camera measurements will measure position and orientation of the target with respect to the camera frame. Hence, the above set of equations are further simplified using small angle approximation. We obtain:

$$\left.\begin{aligned}
\ddot{x} &= (\cos(\phi)\sin(\theta)\cos(\psi) + \sin(\phi)\sin(\psi))\frac{F}{m} \\
\ddot{y} &= (\cos(\phi)\sin(\theta)\sin(\psi) - \sin(\phi)\cos(\psi))\frac{F}{m} \\
\ddot{z} &= (\cos(\phi)\cos(\theta))\frac{F}{m} - g \\
\ddot{\phi} &= \frac{J_y - J_z}{J_x}\dot{\theta}\dot{\psi} + \frac{\tau_\phi}{J_x} \\
\ddot{\theta} &= \frac{J_z - J_x}{J_y}\dot{\phi}\dot{\psi} + \frac{\tau_\theta}{J_y} \\
\ddot{\psi} &= \frac{J_x - J_y}{J_z}\dot{\theta}\dot{\phi} + \frac{\tau_\psi}{J_z}
\end{aligned}\right\} \tag{28}$$

Equation set (28) would be used henceforth for developing control strategies.

4. Vision algorithm development

In order to control a system like the quadrotor, very reliable sensors are needed that can provide a good estimate of the system states. Sensors like the IMU and GPS are subjected to noise which can make them quite undesirable for control applications. Hence, an efficient method of developing control strategies for autonomous quadrotor operations is to utilize the concept of computer-vision.

Using computer vision algorithms, the on-board camera of the quadrotor can be used to confer full autonomy on the system, thereby allowing it to operate in almost any environment. This also eradicates the necessity of setting up an additional set of cameras or to calibrate the environment lighting. As long as the on-board camera is previously calibrated (just needed once) and the target to be tracked is perfectly known (marker size and ID), this system is ready to operate. The usage of ArUco markers as targets allows an easy and fast computation enabling its use in real time applications like autonomous take-off and landing.

In this chapter, let us consider the application of autonomous landing of the quadrotor on a stationary platform like a car roof-top. To enable the quadrotor to identify the landing pad, an ArUco markers board must be attached to the roof of a car. The vision algorithm must be designed to detect a specific ArUco marker ID, and provide the quadrotor's pose relative to the marker. The algorithms used for detection and identification of the marker board are reviewed in the succeeding sub-section.

4.1 The ArUco library

To detect the marker with a regular camera (RGB camera) a library called ArUco is used that was developed by Aplicaciones de la Visión Artificial (AVA) from the Universidad de Córdoba (UCO) [12]. This library is "a minimal library for Augmented Reality applications based on open source computer vision (OpenCV)" [13] and has an API for developing markers in C++ which is very useful in this work. A 100 mm Code 7 ArUco marker is shown in **Figure 2**.

A generic ArUco marker is a 2D bar-code that can be considered as a 7 × 7 Boolean matrix, with the outer cells filled with black (which makes a perfect square, easy to find with image processing). The remaining 5 × 5 matrix is a 10-bits coded ID (up to 1024 different IDs), where each line represents a couple of bits. Each line has only 2 bits of information out of the 5 bits, with the other 3 being used for error detection.

Figure 2.
ArUco marker (ID: 7, size: 100 mm).

Figure 3.
ArUco marker (ID: 1023, size: 100 mm).

These extra 3 bits add asymmetry to the markers, i.e., only a few valid markers are symmetric (e.g., **Figure 3**), which allows a unique orientation detection for the markers. The codification used is a slight modification of the Hamming Code (the first bit is inverted to avoid a valid black square).

So, any ArUco marker can be created by converting a number to binary, splitting into five groups of two bits and by putting each couple in one line of the marker, from the top to the bottom. For example the marker ID of **Figure 2** is the number 7, which is (00 00 00 01 11) in binary. Using the information in **Table 1**, it can be verified that the generated marker is the same as that in **Figure 2**.

The ArUco library processes the image supplied and detects the marker ID as well as its position and orientation in the 3D world, relative to the camera. The open source code of ArUco is based in OpenCV, which is a library highly optimized for

Number	Binary	Marker Code
0	00	
1	01	
2	10	
3	11	

Table 1.
Codification of an ArUco marker.

image processing. Therefore, all the calculations are performed in a matter of seconds so it can be used in real time applications.

The main code is not very complex and the markers detection is performed as follows:

1. Converting color image to gray image.

2. Apply adaptive shareholding.

3. Detect contours.

4. Detect rectangles:

    - Detect corners.

    - Detect linked corners.

    - Consider figures with only four connected corners.

5. For detected markers:

    - Calculate homography (from corners).

    - Threshold the area using OTSU, which assumes a bi-modal distribution and finds the threshold that maximizes the extra-class variance while keeping a low intra-class variance.

    - Detect and identify a valid marker, which respects **Table 1**, and if not detected tries the four rotations.

6. Detect extrinsic parameters (by supplying the calibration matrix, distortion matrix and physical markers dimensions).

The extrinsic parameters are calculated with the help of an OpenCV function: *solvePnP*() [14, 15]. For the marker considered, the four corners of its image and their respective 3D coordinates are provided to the algorithm, which will be:

$$\mathbf{X}_{1-4} = \begin{bmatrix} -d/2 & d/2 & d/2 & -d/2 \\ d/2 & d/2 & -d/2 & -d/2 \\ 0 & 0 & 0 & 0 \end{bmatrix} \tag{29}$$

where $d[m]$ is the dimension of the side of the printed squared marker. As it can be observed, all the four points have their coordinate $Z = 0$ and their (X, Y) coordinates are disposed as a square, which means that the marker is considered to be horizontal, in the origin of the world reference, as suggested in **Figure 4** and so the extrinsic parameters will be the rotation and translation of the camera relative to the marker.

Figure 5 gives the real time implementation of the algorithm for marker detection. The vision algorithm gives the position and orientation offset of the marker center with respect to the camera center. This acts as a reference error to the controller which will use it to position the drone to the center of the ArUco marker and land it. Hence, the succeeding section of this chapter describes the control strategy development.

Figure 4.
A marker at the world's reference.

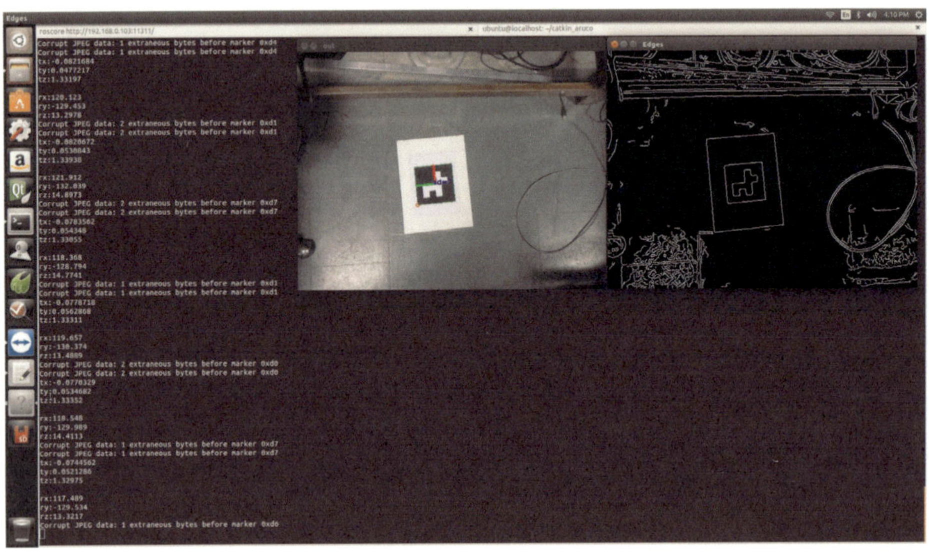

Figure 5.
Real-time implementation.

5. Control strategy

5.1 Introduction to sliding mode control

The sliding mode control (SMC) strategy deals with the design of a sliding manifold also called as a sliding surface which basically describe the desired behavior of the system. The designed control law works to bring the system states onto the user defined sliding surface and then slide them towards the equilibrium point along this surface. The general form of the sliding surface was proposed by Slotine and Li and is defined as [16]:

$$S = \left(\frac{d}{dt} + c_i\right)^{r-1} e = 0 \tag{30}$$

where e is the tracking error defined as $e = x - x_d$. c_i is a positive constant and r is the relative degree of the SMC. In the presence of external disturbances and uncertainties, the system trajectories may deviate from the sliding surface. This can be overcome by making the sliding surface attractive. To ensure sliding surface attractiveness, Lyapunov's theory is utilized as shown below:

Consider the Lyapunov's function V given as:

$$V = \frac{1}{2} S^2 \tag{31}$$

To make the sliding surface attractive and to guarantee asymptotic stability, $\dot{V}$ must be negative definite. In order to make $\dot{V}$ negative definite following condition must be satisfied.

$$S\dot{S} < 0 \tag{32}$$

In order to achieve finite-time convergence (global finite-time stability), the above condition is modified as:

$$S\dot{S} \leq -\eta V^{1/2} \tag{33}$$

To satisfy the above inequality condition, a reaching law is selected as:

$$\dot{S} = -K\,sign(S) \tag{34}$$

where K is the gain and is always positive.

The signum function, $sign(S)$, may be defined as:

$$sign(S) = \begin{cases} +1 & S > 0 \\ 0 & S = 0 \\ -1 & S < 0 \end{cases}$$

The control law generated using SMC has two components defined as [17]:

$$u(t) = u_{eq}(t) + u_h(t) \tag{35}$$

where $u_{eq}(t)$ is the equivalent control, which can be derived by the invariance condition of the sliding surface, i.e., $S = 0$ and $\dot{S} = 0$, and $u_h(t)$ is a hitting control law also called reaching law based control, which can be obtained by testing the attractiveness condition. This hitting law is basically used to overcome the effect of uncertainties and unpredictable disturbances. Chattering appears in SMC due to signum function and can be overcome by using boundary layer method, in which the signum function is replaced by a continuous approximation function like a saturation or hyperbolic function [18].

To understand the basic steps of control law design using sliding mode, Let us consider a second order uncertain nonlinear system [19]

$$\left.\begin{aligned} \dot{x}_1 &= x_2 \\ \dot{x}_2 &= f(x) + g(x)u(t) + d \end{aligned}\right\} \tag{36}$$

where $x = [x_1 \quad x_2]^T$ is the system state vector, $f(x)$ and $g(x) \neq 0$ are smooth nonlinear functions, and bounded uncertain term d satisfies $|d| \leq d_s > 0$, and $u(t)$ is the scalar control input.

Let us define the tracking error as:

$$e = x_1 - x_d \tag{37}$$

where x_d is the desired value of the controlled variable x_1.
The sliding variable is selected as:

$$S = \dot{e} + \lambda e \tag{38}$$

where $\lambda > 0$.
Taking the time derivative of S we get:

$$\begin{aligned} \dot{S} &= \ddot{e} + \lambda \dot{e} \\ &= (\dot{x}_2 - \ddot{x}_d) + \lambda(\dot{x}_1 - \dot{x}_d) \\ &= (f(x) + g(x)u(t) + d - \ddot{x}_d) + \lambda(\dot{x}_1 - \dot{x}_d) \end{aligned} \tag{39}$$

The equivalent control effort which is designed to guarantee desired performance under nominal model is derived as the solution of $\dot{S} = 0$ without considering modeling errors and un-modeled dynamics ($d = 0$). It is represented by u_{eq} and given by:

$$u_{eq} = \frac{1}{g(x)}[\ddot{x}_d - f(x) - \lambda(x_2 - \dot{x}_d)] \tag{40}$$

The hitting control law u_h, to eliminate the effect of perturbations in conventional SMC, is chosen as:

$$u_h = -\frac{K}{g(x)} sign(S) \tag{41}$$

Hence the control law u will be the summation of u_{eq} and u_h and is written as:

$$u = \frac{1}{g(x)}[\ddot{x}_d - f(x) - \lambda(x_2 - \dot{x}_d) - K sign(S)] \tag{42}$$

Now we wish to prove that, for the system Eq. (36), with the sliding variable Eq. (38), if the control law is designed as:

$$u = \frac{1}{g(x)}[\ddot{x}_d - f(x) - \lambda(x_2 - \dot{x}_d) - K sign(S)] \tag{43}$$

with $\lambda > 0$ then the $S = 0$ will be reached in finite time. Also, the states x_1 and x_2 will converge to zero asymptotically. We use the Lyapunov's stability criteria: Let us choose the following Lyapunov candidate function as:

$$V = \frac{1}{2}S^2 \tag{44}$$

Taking the time derivative of V we get:

$$\begin{aligned} \dot{V} &= S\dot{S} \\ &= S[f(x) + g(x)u + d - \ddot{x}_d + \lambda(x_2 - \dot{x}_d)] \end{aligned} \tag{45}$$

From Eq. (42), Substitute u in Eq. (45) then

$$\begin{aligned}\dot{V} &= S(-K\,sign(S) + d) \\ &\leq -K|S| + d_s|S| \\ &= -|S|(K - d_s) \leq -\frac{\eta}{\sqrt{2}}|S|\end{aligned} \tag{46}$$

Since $\dot{V}$ is negative semi definite for $K \geq d_s + \frac{\eta}{\sqrt{2}}$. This ensure the finite time convergence of the sliding manifold. As a result, states are converging to desired value asymptotically. There are two phases associated with sliding mode control namely reaching phase and sliding phase. The reaching phase, is the part where the state trajectory starts from its initial condition and moves toward the sliding surface. In sliding phase, trajectories moves only on the desired sliding surface. The time taken by the states to reach sliding surface is called reaching time, denoted as t_r. To derive an expression for t_r: From Eq. (33), we can write

$$\dot{V} = -\eta V^{1/2} \tag{47}$$

Indeed, separating variables and integrating Eq. (47) over the time interval $0 \leq t \leq t_r$, we obtain

$$t_r = \frac{2}{\eta} V(0)^{1/2} \tag{48}$$

Therefore, a control u that is computed to satisfy Eq. (47) will drive the variable S to zero in finite time t_r and will keep it at zero thereafter. Now we extend this idea to the quadrotor by using the model represented by the equation set (28).

5.2 SMC design for quadrotor

Let us represent the non-linear model of the quadrotor as:

$$\dot{\mathbf{x}} = \mathbf{f}(\mathbf{x}) + \mathbf{g}(\mathbf{x})\mathbf{u} + \mathbf{d} \tag{49}$$

where:

$$\mathbf{f}(\mathbf{x}) = \begin{pmatrix} \dot{\phi} \\ \left(\frac{J_y - J_z}{J_x}\right)\dot{\theta}\dot{\psi} \\ \dot{\theta} \\ \left(\frac{J_z - J_x}{J_y}\right)\dot{\phi}\dot{\psi} \\ \dot{\psi} \\ \left(\frac{J_x - J_y}{J_z}\right)\dot{\phi}\dot{\theta} \\ \dot{z} \\ -g \\ \dot{x} \\ 0 \\ \dot{y} \\ 0 \end{pmatrix}, \mathbf{d} = \begin{pmatrix} 0 \\ d_\phi \\ 0 \\ d_\theta \\ 0 \\ d_\psi \\ 0 \\ d_z \\ 0 \\ d_x \\ 0 \\ d_y \end{pmatrix} \tag{50}$$

and

$$\mathbf{g}(\mathbf{x}) = \begin{pmatrix} 0 & 0 & 0 & 0 \\ 0 & 1/J_x & 0 & 0 \\ 0 & 0 & 0 & 0 \\ 0 & 0 & 1/J_y & 0 \\ 0 & 0 & 0 & 0 \\ 0 & 0 & 0 & 1/J_z \\ 0 & 0 & 0 & 0 \\ (\cos\phi\cos\theta)/m & 0 & 0 & 0 \\ 0 & 0 & 0 & 0 \\ u_x/m & 0 & 0 & 0 \\ 0 & 0 & 0 & 0 \\ u_y/m & 0 & 0 & 0 \end{pmatrix} \tag{51}$$

Here, the terms u_x and u_y are termed as virtual inputs and are evaluated as:

$$\begin{aligned} u_x &= \cos\phi\sin\theta\cos\psi + \sin\phi\sin\psi \\ u_y &= \cos\phi\sin\theta\sin\psi - \sin\phi\cos\psi \end{aligned} \tag{52}$$

From Eq. (49), the state vector can be expressed as $\mathbf{x} = \left(\phi, \dot{\phi}, \theta, \dot{\theta}, \psi, \dot{\psi}, z, \dot{z}, x, \dot{x}, y, \dot{y}\right)^T$ and the control input vector as $\mathbf{u} = (u_1, u_2, u_3, u_4)^T$ which corresponds to $\left(F, \tau_\phi, \tau_\theta, \tau_\psi\right)$. $\mathbf{d}$ represents bounded lumped disturbance which is a sum of modeling uncertainties and external wind gust disturbance associated with the quadrotor dynamics. For convenience, let the states of the system be renamed as: $\mathbf{x} = (x_1, x_2, x_3, x_4, x_5, x_6, x_7, x_8, x_9, x_{10}, x_{11}, x_{12})^T$.

The quadrotor dynamical model can be split into 6 second-order sub-systems, namely the altitude, x-position, y-position, roll, pitch and yaw sub-systems. The altitude and yaw sub-systems are controlled directly by u_1 and u_4. However, the position sub-systems are coupled with the roll and pitch sub-systems. Hence, the concept of virtual control is utilized to develop the control scheme. Hence, u_x and u_y will control the x and y positions and u_2 and u_3 will control the roll and pitch sub-systems.

In order to design u_2, let us consider the roll subsystem which can be obtained from Eq. (49) given as:

$$\left.\begin{aligned} \dot{x}_1 &= x_2 \\ \dot{x}_2 &= \frac{J_y - J_z}{J_x} x_4 x_6 + \frac{u_2}{J_x} + d_\phi \end{aligned}\right\} \tag{53}$$

As mentioned previously, d_ϕ is the bounded lumped uncertainty in the roll dynamics with an upper bound of d_s. Let us consider the tracking error in roll angle as:

$$e_\phi = x_1 - \phi_d \tag{54}$$

where ϕ_d is computed from Eq. (52) as:

$$\phi_d = \sin^{-1}\left(u_x \sin\psi_d - u_y \cos\psi_d\right) \tag{55}$$

The sliding variable is defined as:

$$S_\phi = \dot{e}_\phi + \lambda_1 e_\phi \tag{56}$$

Taking the time derivative of S_ϕ:

$$\begin{aligned} \dot{S}_\phi &= \ddot{e}_\phi + \lambda_1 \dot{e}_\phi \\ &= \left(\ddot{\phi} - \ddot{\phi}_d\right) + \lambda_1\left(x_2 - \dot{\phi}_d\right) \end{aligned} \tag{57}$$

In order to eliminate the disturbance effects, the reaching law is selected as:

$$\dot{S}_\phi = -K_1 sign\left(S_\phi\right) \tag{58}$$

From Eqs. (57) and (58) the control law can be chosen as:

$$u_2 = J_x\left[\ddot{\phi}_d - \frac{J_y - J_z}{J_x} x_4 x_6 - \lambda_1\left(x_2 - \dot{\phi}_d\right) - K_1 sign\left(S_\phi\right)\right] \tag{59}$$

On similar lines, u_1, u_x, u_y, u_3 and u_4 are designed as:

$$u_3 = J_y\left[\ddot{\theta}_d - \frac{J_z - J_x}{J_y} x_2 x_6 - \lambda_2\left(x_4 - \dot{\theta}_d\right) - K_2 sign\left(S_\theta\right)\right] \tag{60}$$

where θ_d is computed from 52 and is given as:

$$\theta_d = \sin^{-1}\left(\frac{u_x \cos\psi_d + u_y \sin\psi_d}{\sqrt{1 - \left(u_x \sin\psi_d - u_y \cos\psi_d\right)^2}}\right) \tag{61}$$

$$u_4 = J_z\left[\ddot{\psi}_d - \frac{J_x - J_y}{J_z} x_2 x_4 \quad \lambda_3\left(x_6 - \dot{\psi}_d\right) - K_3 sign\left(S_\psi\right)\right] \tag{62}$$

$$u_1 = \frac{m}{\cos x_1 \cos x_3}\left[\ddot{z}_d + g - \lambda_4\left(x_8 - \dot{z}_d\right) - K_4 sign\left(S_z\right)\right] \tag{63}$$

The sliding variables are expressed as:

$$\left.\begin{aligned} S_\theta &= \dot{e}_\theta + \lambda_2 e_\theta \\ S_\psi &= \dot{e}_\psi + \lambda_3 e_\psi \\ S_z &= \dot{e}_z + \lambda_4 e_z \end{aligned}\right\} \tag{64}$$

With the tracking errors as:

$$\left.\begin{aligned} e_\theta &= x_3 - \theta_d; \\ e_\psi &= x_5 - \psi_d; \\ e_z &= x_7 - z_d; \end{aligned}\right\} \tag{65}$$

To achieve x and y motion control, the virtual inputs u_x and u_y are designed as:

$$\begin{aligned} u_x &= \frac{m}{u_1}\left[\ddot{x}_d - \lambda_5\left(x_{10} - \dot{x}_d\right) - K_5 sign\left(S_x\right)\right] \\ u_y &= \frac{m}{u_1}\left[\ddot{y}_d - \lambda_6\left(x_{12} - \dot{y}_d\right) - K_6 sign\left(S_y\right)\right] \end{aligned} \tag{66}$$

where

$$\left.\begin{aligned} S_x &= \dot{e}_x + \lambda_5 e_x \\ S_y &= \dot{e}_y + \lambda_6 e_y \end{aligned}\right\} \tag{67}$$

and

$$\left.\begin{aligned} e_x &= x_9 - x_d; \\ e_y &= x_{11} - y_d; \end{aligned}\right\} \tag{68}$$

As previously done, the task now is to prove that the system Eq. (49), with the sliding variables given by Eqs. (56), (64) and (67). If the control laws are designed as Eqs. (59), (60), (62), (63), and (66) then the sliding manifolds are reached in finite time t_r and the tracking error e_ϕ, e_θ, e_ψ, e_z, e_x, e_y will stay on the sliding manifolds thereafter. Consequently the controlled states $x_1, x_3, x_5, x_7, x_9, x_{11}$ will converge to the desired values in finite time t_f in the presence of bounded disturbance and uncertainties.

To do so, let us select a candidate Lyapunov function as:

$$V = \frac{1}{2}\mathbf{S}^{\mathbf{T}}\mathbf{S} \tag{69}$$

where $\mathbf{S} = \left(S_\phi, S_\theta, S_\psi, S_z, S_x, S_y\right)^T$. By taking the time derivative of the Lyapunov energy function Eq. (69), one can get:

$$\dot{V} = \mathbf{S}^{\mathbf{T}}\dot{\mathbf{S}} \tag{70}$$

where $\dot{\mathbf{S}} = \left(\dot{S}_\phi, \dot{S}_\theta, \dot{S}_\psi, \dot{S}_z, \dot{S}_x, \dot{S}_y\right)^T$. After substitution of S, $\dot{V}$ can be expressed as:

$$\dot{V} = \mathbf{S}^{\mathbf{T}}(\ddot{\mathbf{e}} + A\dot{\mathbf{e}}) \tag{71}$$

where $\dot{\mathbf{e}} = \left(\dot{e}_\phi, \dot{e}_\theta, \dot{e}_\psi, \dot{e}_z, \dot{e}_x, \dot{e}_y\right)^T$ and $\mathbf{A}$ is the diagonal matrices where $\mathbf{A} = diag\{\lambda_1, \lambda_2, \lambda_3, \lambda_4, \lambda_5, \lambda_6\}$ with $\lambda_i > 0$. Substituting the value of designed control laws in Eq. (71):

$$\dot{V} = \mathbf{S}^{\mathbf{T}}(-\mathbf{Ksign}(\mathbf{S}) + \mathbf{d}) \tag{72}$$

where $\mathbf{K} = diag\{K_1, K_2, K_3, K_4, K_5, K_6\}$ with $K_i > 0$ and $\mathbf{d} = \left(d_\phi, d_\theta, d_\psi, d_z, d_x, d_y\right)^T$ and $sign(\mathbf{S}) = \left(sign\left(S_\phi\right), sign(S_\theta), sign\left(S_\psi\right), sign(S_z), sign(S_x), sign\left(S_y\right)\right)^T$

$$\begin{aligned} \dot{V} &= \mathbf{S}^{\mathbf{T}}(-\mathbf{Ksign}(\mathbf{S}) + \mathbf{d}) \\ &\le \sum_{i=1}^{6}(|S_i|d_{si} - K_i|S_i|) \\ &= -\sum_{i=1}^{6}|S_i|(-d_{si} + K_i) \\ &\le -\frac{\eta_1}{\sqrt{2}}\sum_{i=1}^{6}|S_i| \end{aligned} \tag{73}$$

where $|d_i| < d_{si}$. Hence, the convergence of **S** is proven by the Lyapunov stability theory. The sliding variables are converging to zero in finite time, i.e., $\mathbf{S} \to 0$. Therefore, tracking error will converge to zero asymptotically, i.e., $\mathbf{e} \to 0$.

6. Simulation results

This section presents the simulation results of the SMC described in the previous section. The tracking performance of the quadrotor is evaluated by making it track a circle of radius 1 m at an altitude of 3 m with a desired yaw angle of $\pi/6$. The tracking performance is shown in **Figures 6–9**.

The control inputs are shown in **Figures 10–13**. One can observe that there exists a presence of chattering in the control inputs when using the signum function and cannot be directly implemented in real-time hardware. To overcome this, the boundary layer approximation is utilized which smoothens the control inputs.

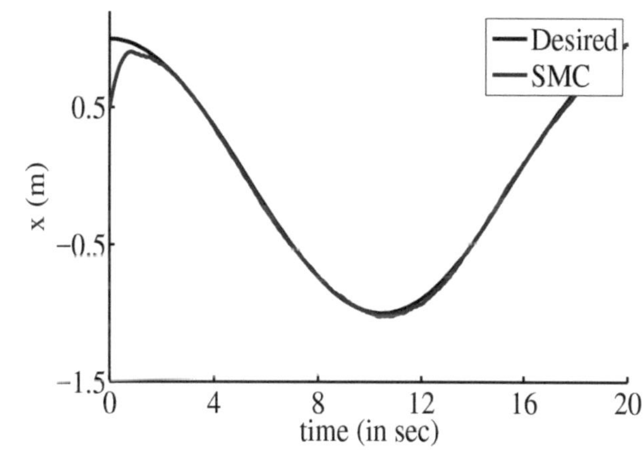

Figure 6.
X-position tracking.

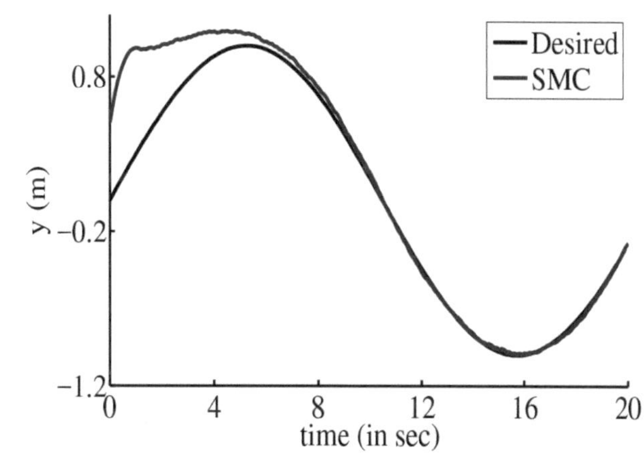

Figure 7.
Y-position tracking.

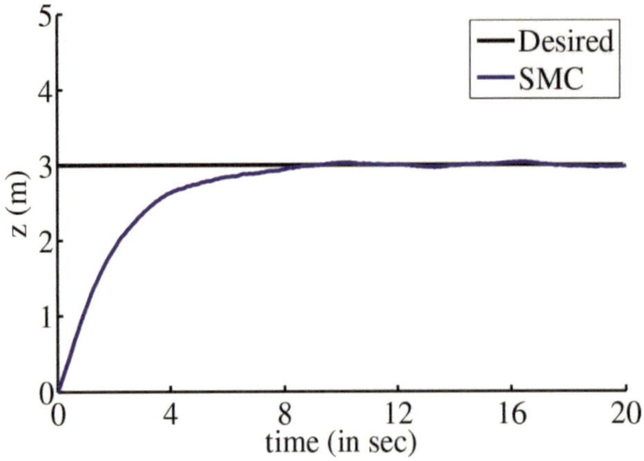

Figure 8.
Altitude tracking.

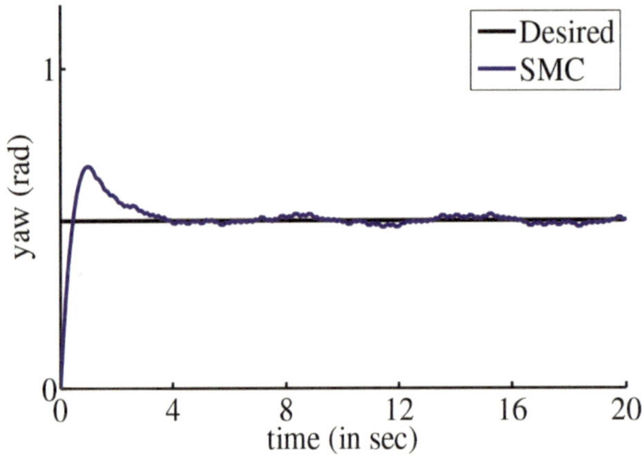

Figure 9.
Yaw tracking.

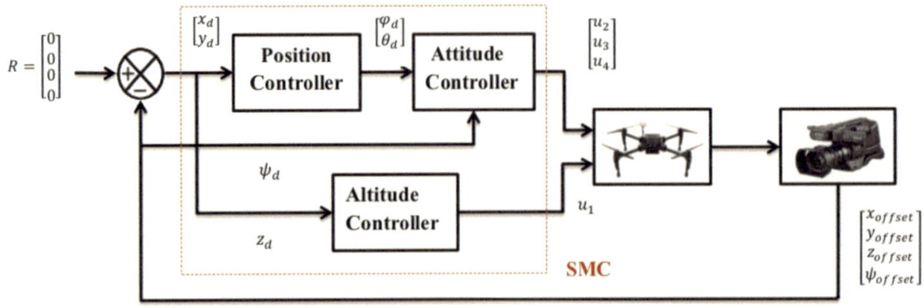

Figure 10.
u_1.

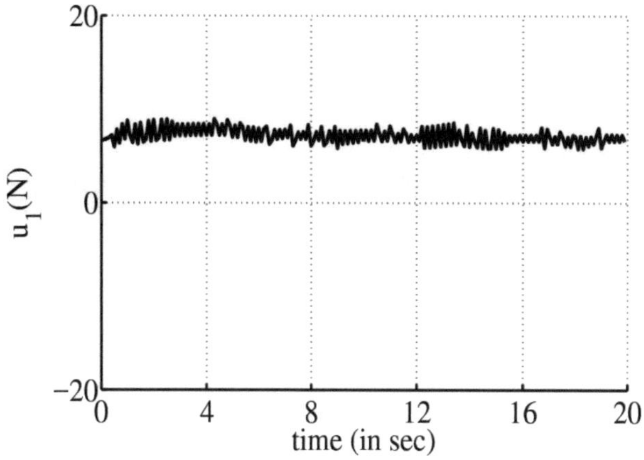

Figure 11.
u_2.

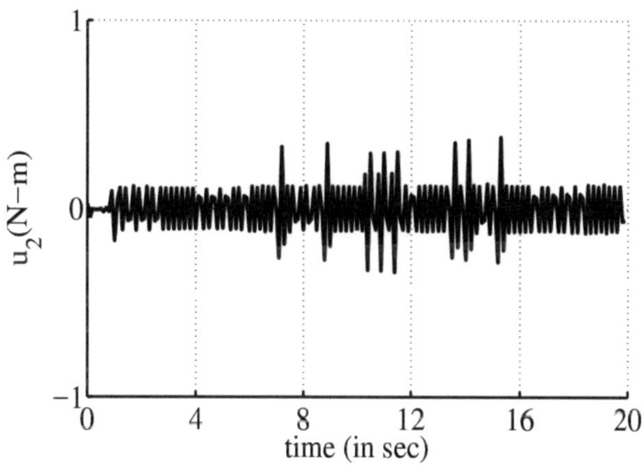

Figure 12.
u_3.

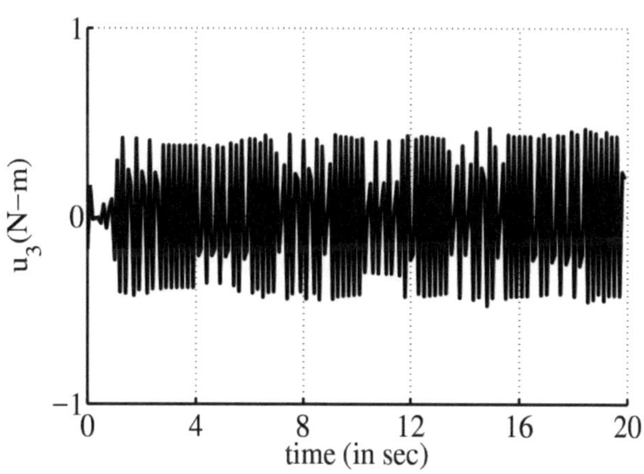

Figure 13.
u_4.

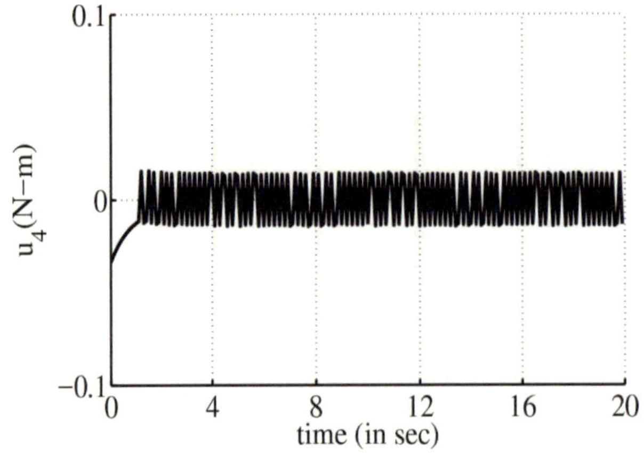

Figure 14.
Complete block diagram.

7. Vision-integrated control

Figure 14 presents an overview of the vision-integrated control. The camera captures the image and based on the vision algorithm, the position and orientation offset between the quadrotor and the marker is obtained. This offset is fed to the sliding mode controller which reduces this error and aids in landing the quadrotor at the center of the marker.

8. Hardware results

This section presents the results obtained from real-time implementation of the vision-integrated sliding mode control for the autonomous landing of the quadrotor in indoor and outdoor environments.

8.1 Hardware description

As mentioned in earlier parts of this chapter, the quadrotor used for the implementation of this work is the *DJI Matrice M100* shown in **Figure 15**. The DJI Matrice

Figure 15.
DJI Matrice M100.

100 is a fully customize-able and programmable flight platform that lets its users perform operations such as pipeline health monitoring, surveillance, search and rescue and in applications requiring external sensor interface. Accompanied with the M100, a series of add-ons help in making its handling user-friendly. Similar to any other development drone in the market, the Matrice M100 comes with a programmed flight controller.

To aid in implementation of user defined controllers and task maneuvers, a separate on-board computer, named the *DJI Manifold*, is provided in **Figure 16**. The Manifold is an embedded Linux computer which incorporates a NVIDIA Tegra K1 SOC (CPU + GPU + ISP in a single chip) with both standard and extended connections and interfaces. The single GPU (Graphical Processing Unit) unit helps us run CUDA to aid in performing complex image processing operations. The Linux environment acts as a support to run ROS (Robot Operating System), which is the key element for any sorts of development on the Matrice M100. This would be mentioned in detail in the upcoming sub-section.

To gather visual data, the DJI Matrice M100 is provided with a completely controllable Zenmuse X3 Gimbal. This could be easily interfaces with the DJI Manifold for image processing. However, in this case, a separate downward facing camera is used to perform the task of vision based landing. This is done so as to keep the gimbal free to perform other tasks such as image capturing, video capturing and likewise. The downward facing camera chosen is the *LogiTech C310* camera (**Figure 17**) which can be interfaced with the manifold using an USB connection.

The landing pad is a wooden platform of dimension 4 feet × 4 feet. At the center, an AruCo marker is placed of dimension 12.5 cm × 12.5 cm. The AruCo Marker chosen is a 4 × 4 matrix of marker ID 7. The dimension of the marker is chosen such that it is clearly detected from an altitude as high as 10 m as well as from an altitude as low as 0.4 m. The landing pad setup as shown in **Figure 18** would be mounted on the roof of a car for experimental purposes.

8.2 Software description

This section briefly describes the software abstraction layer and its paradigm to control and the associated hardware flow of Matrice M100 quadrotor. As discussed in the hardware setup the DJI M100 uses DJI Manifold as its on-board computer to control and communicate with Flight controller and on-board sensors interfaced with it. *DJI On-board SDK (OSDK)* is an open source software library which enables the OBC (On-Board Computer) to handle the Input-Output data coming from the

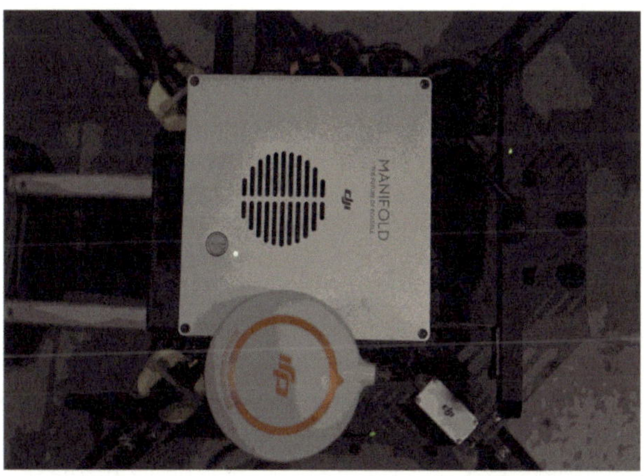

Figure 16.
DJI manifold.

Figure 17.
LogiTech C310.

Figure 18.
Landing pad setup.

on-board control unit and sensor units. To establish the reliable network among the on-board sensor units and OBC, several serial communication protocols such as *UART1, UART2, CAN1, CAN2, USB* and *VBUS1 to VBUS5* are used. In this Paper, the main focus is on estimating the pose of the quadrotor using an on-board monocular camera connected to one of the USB ports. Other sensors, such as the *DJI Guidance*, which is connected to the *VBUS*, can be sued for fusion at different frame rates if necessary. The multi-layer hardware communication block diagram is as shown in **Figure 19**.

The multi-layer hardware connection is described in the **Figure 19**.

The on-board SDK includes:

- C++ library to access arm processor based linux(OS).
- Robot Operating System *(ROS)*: Interface and associated packages to handle multiple sensor nodes.
- *DJI Assistant2*: Real time flight simulator to verify the developed algorithms.
- *DJI OSDK API*: Used to asynchronously to send the control commands to flight control unit and s the acknowledgment from it.

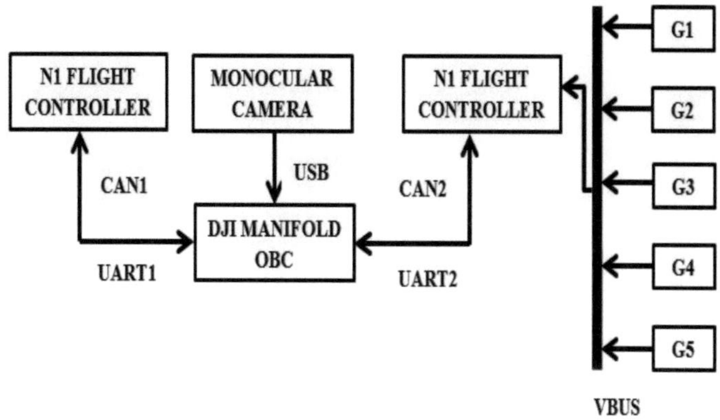

Figure 19.
OSDK architecture and software framework.

The software components of OSDK consist of APIs provided by DJI SDK library. The OSDK supports two varieties of asynchronous Programming and sends information to the OSDK workflow. The asynchronous programming mechanism works on executing the code receiving from the acknowledgement which is independent of main flow execution. The components also include:

- Serial device drivers: It communicates with flight controller and OBC via *UART*. The serial device drivers also takes care of input-output handling, memory management like locking and unlocking and interrupts.

- Thread communication: Allows inter thread communication to handle different level of signals.

- Application layer API calls: The core of on-board API is a communication between the flight control commands send from the processor to the control unit and in turn receives the acknowledgement independent of program flow. It provides callback functions. The synchronous programming API blocking calls will return only when the CMD-ACK round trip is done. This gives the assurance that the command is executed.

This process flow is depicted as shown in **Figure 20**.

8.3 Test environment description

Two test environments were used to validate the developed control algorithm. To assess the quadrotor's capability of performing vision based landing in the indoor environment, an empty plot of dimension 12 feet × 21 feet was used enclosed by nets. The plot was surrounded with obstacles on all four sides making it absolutely necessary for the drone not to move away too far away from the landing pad. The test environment is as shown in **Figure 21**. The first set of experiments were conducted using this setup. This also gave an opportunity to validate the shadow elimination that was incorporated in the drone. Note that the indoor experiment had the landing pad setup placed on the ground.

The second setup included the landing pad placed on the roof of a car as shown in **Figure 22**. It is assumed that the car is stationary when the quadrotor is performing the task of vision based landing.

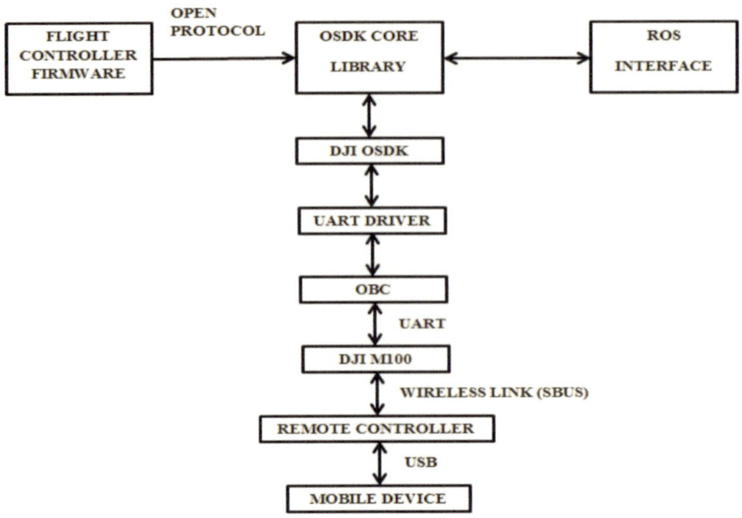

Figure 20.
OSDK software flow.

Figure 21.
Indoor test environment.

8.4 Results

8.4.1 Indoor environment

The first environment was the indoor environment with the marker placed on the ground in an enclosed space of 12 feet × 12 feet. The drone was made to autonomously lift-off and then the vision based landing node was initiated. The node simultaneously also recorded the position, velocity, acceleration and offset as the drone performed the corrections needed to align with the marker.

These results were plotted and are shown in **Figures 23–26**. Note that 1 s produces 18 samples and hence, from the time of initiation to completion, the action of

Figure 22.
Outdoor test environment.

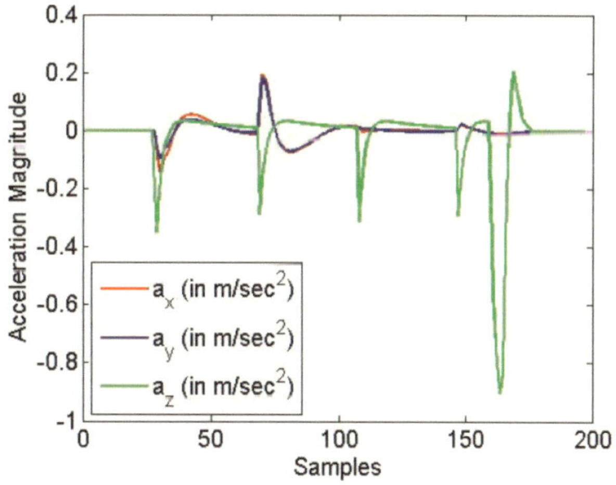

Figure 23.
Acceleration profile (Indoor Testing).

vision based landing took 30 s in this case. Over 10 trials an average error of 3.2 cm was observed with the maximum error as 6 cm from the marker center.

8.4.2 Outdoor environment

The second test environment was the outdoor environment with the landing pad mounted on the roof of a car. A 4 $\times$ 4 feet wooden board was mounted on the roof top of a car with the ArUco marker affixed to the center of this board. It was tested in an open ground with winds blowing at 10 km/hr. NW. This helped us understand the robustness of the controller designed. A slight swaying of the drone was observed, however, the designed controller managed to land the quadrotor on the marker with an average error of 4 cm with a maximum error of 7 cm over 20 trials. Once again, the acceleration, velocity, position and offset values were recorded and

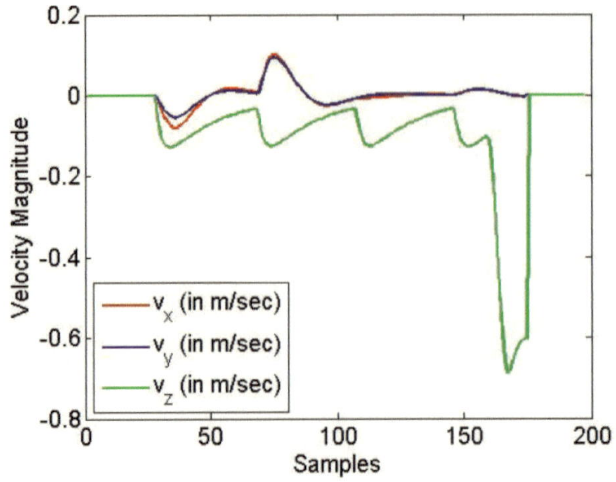

Figure 24.
Velocity profile (Indoor Testing).

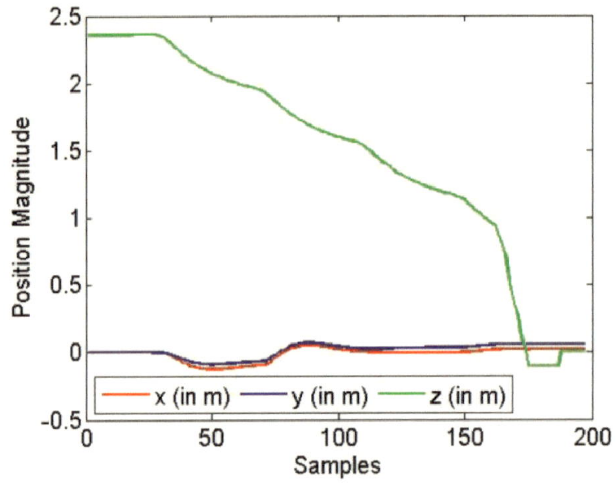

Figure 25.
Position profile (Indoor Testing).

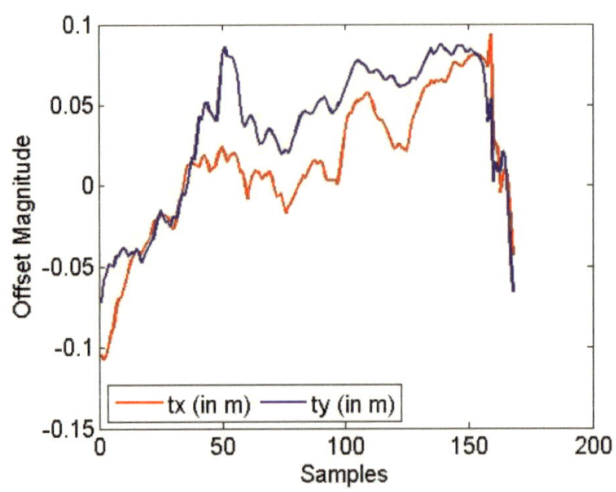

Figure 26.
Camera parameters (Indoor Testing).

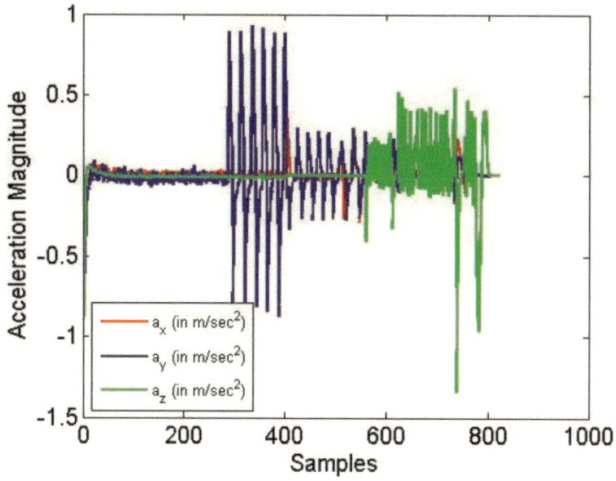

Figure 27.
Acceleration profile (Outdoor Testing).

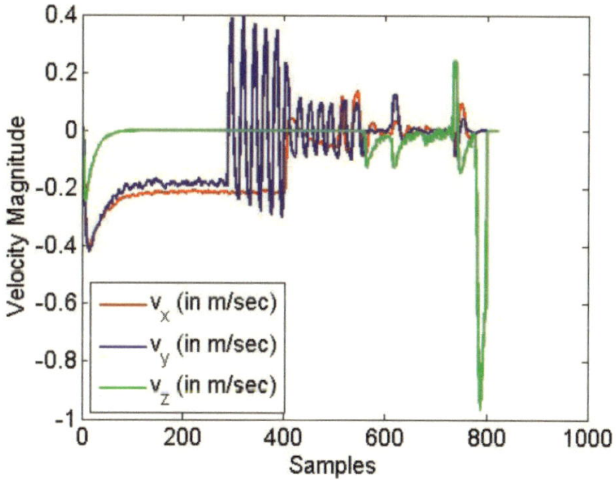

Figure 28.
Velocity profile (Outdoor Testing).

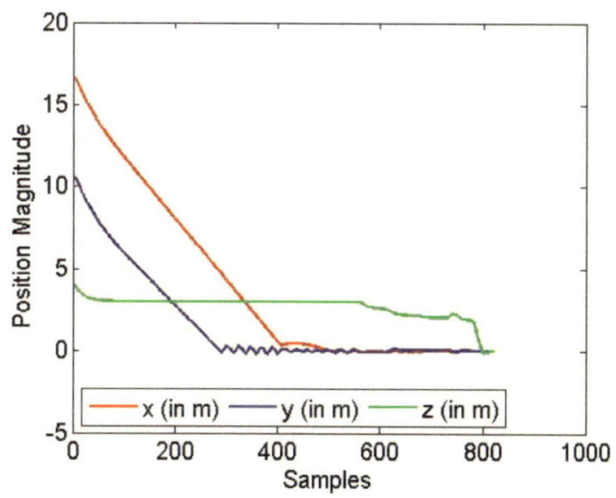

Figure 29.
Position profile (Outdoor Testing).

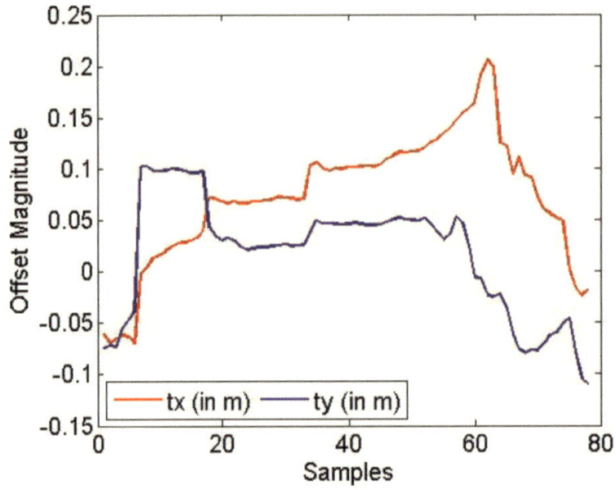

Figure 30.
Camera parameters (Outdoor Testing).

are shown in the **Figures 27–30**. The time for completion of the task from the point of initialization was found to be 43 s.

9. Conclusion

In this work, a vision-based sliding mode control for autonomous landing of a quadrotor UAV is proposed. The vision algorithm is developed to detect the centroid, position and orientation of the camera with respect to a landing pad marker (ArUco marker) placed on the roof of a car. The designed sliding mode controller proves to be effective when working alongside the developed vision algorithm and is simulated using MATLAB environment. This is then on extended to the actual experimental tests on the DJI Matrice M100, in indoor and outdoor environments. The main conclusions are summarized as follows:

1. The designed controller ensures that all the state variables converge to their reference values, even if their reference values are subjected to sudden changes.

2. The alignment of the drone over the landing pad marker is obtained by using the position and yaw offset values as inputs to the sliding mode controller.

3. The robustness of the designed controller is demonstrated among the various experimental trials in outdoor environments (subjected to winds), and the effectiveness of the proposed control scheme is also justified.

All of the results presented above are quiet promising and can be reproduced in any quadrotor system. Reference [20] demonstrates the results of the proposed work. As a future addition to this work, readers can consider using EKF to infuse IMU data with vision to enhance the tracking data. In addition, the users can also improve the proposed SMC to incorporate power rate reaching laws or super twisting laws to attenuate chattering further.

Acknowledgements

The authors would like to thank the Subhash Chand Yogi and Abhay Pratap Singh from Indian Institute of Technology-Kanpur, who have helped in running simulations and hardware experiments.

Author details

Archit Krishna Kamath, Vibhu Kumar Tripathi and Laxmidhar Behera*
Department of Electrical Engineering, Indian Institute of Technology, Kanpur, India

*Address all correspondence to: lbehera@iitk.ac.in

References

[1] Besnard L, Shtessel YB, Landrum B. Quadrotor vehicle control via sliding mode controller driven by sliding mode disturbance observer. Journal of the Franklin Institute. 2012;**349**(2):658-684

[2] Zheng EH, Xiong JJ, Luo JL. Second order sliding mode control for a quadrotor UAV. ISA Transactions. 2014; **53**(4):1350-1356

[3] Bouchoucha M, Seghour S, Tadjine M. Classical and second order sliding mode control solution to an attitude stabilization of a four rotors helicopter: From theory to experiment. In: 2011 IEEE International Conference on Mechatronics; IEEE; 2011. pp. 162-169

[4] Xu R, Özgüner Ü. Sliding mode control of a class of underactuated systems. Automatica. 2008;**44**(1): 233-241

[5] Mokhtari A, Benallegue A, Orlov Y. Exact linearization and sliding mode observer for a quadrotor unmanned aerial vehicle. International Journal of Robotics and Automation. 2006;**21**(1): 39-49

[6] Benallegue A, Mokhtari A, Fridman L. High-order sliding-mode observer for a quadrotor UAV. International Journal of Robust and Nonlinear Control: IFAC-Affiliated Journal. 2008;**18**(4-5):427-440

[7] Sharifi F, Mirzaei M, Gordon BW, Zhang Y. Fault tolerant control of a quadrotor UAV using sliding mode control. In: 2010 Conference on Control and Fault-Tolerant Systems (SysTol), IEEE; 2010. pp. 239-244

[8] Ashrafiuon H, Erwin RS. Sliding mode control of underactuated multibody systems and its application to shape change control. International Journal of Control. 2008;**81**(12): 1849-1858

[9] Tripathi VK, Behera L, Verma N. Design of sliding mode and backstepping controllers for a quadcopter. In: 2015 39th National Systems Conference (NSC), IEEE; 2015. pp. 1-6

[10] Erginer B, Altug E. Modeling and PD control of a quadrotor VTOL vehicle. In: 2007 IEEE Intelligent Vehicles Symposium, IEEE; 2007. pp. 894-899

[11] Beard R. Quadrotor dynamics and control rev 0.1

[12] Garrido-Jurado S, Muñoz-Salinas R, Madrid-Cuevas FJ, Marín-Jiménez MJ. Automatic generation and detection of highly reliable fiducial markers under occlusion. Pattern Recognition. 2014;**47** (6):2280-2292

[13] Bradski G, Kaehler A. Learning OpenCV: Computer vision with the OpenCV Library. "O'Reilly Media, Inc."; 2008

[14] Kannala J, Brandt SS. A generic camera model and calibration method for conventional, wide-angle, and fish-eye lenses. IEEE Transactions on Pattern Analysis and Machine Intelligence. 2006;**28**(8):1335-1340

[15] Munoz-Salinas R. Aruco: A minimal library for augmented reality applications based on opencv. Universidad de Córdoba; 2012

[16] Slotine JJ, Li W. Applied Nonlinear Control. Englewood Cliffs, NJ: Prentice Hall; 1991

[17] Shtessel Y, Edwards C, Fridman L, Levant A. Sliding Mode Control and Observation. New York: Springer; 2014

[18] Runcharoon K, Srichatrapimuk V. Sliding mode control of quadrotor. In:

2013 The International Conference on Technological Advances in Electrical, Electronics and Computer Engineering (TAEECE), IEEE; 2013. pp. 552-557

[19] Kuo BC, Golnaraghi F. Automatic Control Systems. Englewood Cliffs, NJ: Prentice-Hall; 1995

[20] Kamath AK, Sakthi Vignesh R, Behera L. Vision Based Autonomous Landing on Stationary Platform. [Online]. Available from: https://www.youtube.com/watch?v=d-D7enlMzlo

Chapter 4

A System for Continuous Underground Site Mapping and Exploration

Alexander Ferrein, Ingrid Scholl, Tobias Neumann, Kai Krückel and Stefan Schiffer

Abstract

3D mapping becomes ever more important not only in industrial mobile robotic applications for AGV and production vehicles but also for search and rescue scenarios. In this chapter we report on our work of mapping and exploring underground mines. Our contribution is two-fold: First, we present our custom-built 3D laser range platform SWAP and compare it against an architectural laser scanner. The advantages are that the mapping vehicle can scan in a continuous mode and does not have to do stop-and-go scanning. The second contribution is the mapping tool mapit which supports and automates the registration of large sets of point clouds. The idea behind mapit is to keep the raw point cloud data as a basis for any map generation and only store all operations executed on the point clouds. This way the initial data do not get lost, and improvements on low-level date (e.g. improved transforms through loop closure) will automatically improve the final maps. Finally, we also present methods for visualization and interactive exploration of such maps.

Keywords: 3D mapping, continuous mapping, large underground site mapping, mapping tools, point cloud registration, map exploration, map visualization

1. Introduction

An important environment information in urban search and rescue applications is 3D map data of the site. First responders usually have 2D map material of buildings, tunnels and street sites, while a concise overview in 3D would possibly give them further and new important information about the situation at hand. Kruijff et al. [1, 2], for instance, report from mapping an earthquake site in Mirandola, Italy, in July 2012. There, an unmanned guided vehicle (UGV) together with unmanned aerial vehicle (UAV) acquired 3D environment information to help judge if partly destroyed structures are safe for first responders to move in. Also, robotic technology may help reveal information that is inaccessible otherwise. For example, [3] present an approach for UAV-based mapping of underground tunnels in darkness. While more works focus on disaster management with UAVs (e.g. [4]), many others build upon ground-based USAR robots with the capability to map the disaster site in 3D (see, for instance [5, 6]). Many other related research works including our own previous research [7–9] focus on investigating SLAM algorithms best suited for mapping disaster sites. How large outdoor environments can be mapped in a fast manner is investigated in [10]. A general overview on robotics in disaster scenarios can be found, for example, in [11]. Again other work looks at sensor systems [12] that are appropriate for generating feasible maps. As shown in [13], for first responders it is important to get a reliable overview of the disaster site and the damages and hazards in order to react correctly. This motivates the work presented in this paper. It is very important to get a map quite quickly in order to judge the operation on-site. The operation may be inspecting a disaster site, mapping a building site or, as in our case, mapping an underground mining operation: in each of these examples, the operator needs to quickly get an overview of the site. The operator then either learns where further sensor information needs to be acquired or what measures need to be taken for the operation. In this paper, we report on our 3D mapping system which offers exactly this. In a slightly different but comparable setting in underground mines, we developed a mapping system in hardware and software which allows to quickly integrate new laser scans into a 3D map.

1.1 The project UPNS4D+

The results presented in this paper are part of the project "Underground 4D+ Positioning, Navigation and Mapping System for Highly Selective, Efficient and Highly-secure Exploitation of Important Resources" (UPNS4D+) which was funded by the German Federal Ministry of Education and Research within the programme of "R4–Innovative Technologies for Resource Efficiency – Research for the Provision of Raw Materials of Strategic Economic Importance".

The overall project aimed at exploiting mineral resources of rare earths in a highly selective, efficient and highly secure way from local deposits as well as detecting new ones. This required innovative mining technologies which integrate dynamic change of the mine. The interdisciplinary research project UPNS4D+ aimed at developing an underground deposit positioning, navigation and mapping system for a mobile robot platform. For more details on the overall project, we refer to [14, 15].

The consortium consisted of the following partners: (1) indurad GmbH, Aachen; (2) Fachhochschule Aachen, MASCOR Institute; (3) MILAN Geoservice GmbH, Spremberg; (4) RWTH Aachen University, Institute for Advanced Mining Technology; (5) XGraphic Ingenieurgesellschaft mbH, Aachen; (6) Technische Universität Bergakademie Freiberg; (7) Fritz Rensmann, Maschinenfabrik, Diesellokomotiven, Getriebe GmbH & Co. KG, Dortmund; and (8) GHH Fahrzeuge GmbH, Gelsenkirchen. The project started in April 2015 and ended in December 2018.

The goal of our subproject "6D mapping" was to develop a prototype robot system that is able to map underground mining sites. To this end, a suitable robot platform had to be equipped with the right sensor equipment (radar, cameras, LiDARs, IMU). The data coming from the sensors needed to be integrated into consistent high-dimensional maps deploying known SLAM approaches. High-dimensional means that besides the 3D point clouds, also key frames from the vision or radar data were stored in the map. New approaches had to be developed to grant easy access to the data in order to process and visualize them.

1.2 Contribution

In the following, we present results from that research project that are highly relevant also for urban search and rescue robotics and will find useful applications there. In particular, we present:

1. A novel sensor platform [16] which allows for continuous high-resolution scans

2. A novel registration tool for checking point clouds on-site

3. The sensor data registration tool mapit which facilitates the processing of large-scale point cloud data.

1.3 Outline

The paper is organized as follows. In the next section (Section 2), we present the exploration vehicle that was developed during the project and used in our experiments. We introduce the overall platform design and the sensor setup which was used for exploration runs at underground mining site. Our exploration robot is equipped with a revolving 3D LiDAR for acquiring map data, several further 3D and 2D laser range finders used for navigation and terrain classification, a thermal imaging camera for detecting mine workers even in unilluminated areas and a high-resolution wide-angle camera for teleoperation. Additionally, we mounted a FARO 3D LiDAR as a reference system.

Section 3 is devoted to our novel 3D LiDAR platform SWAP. SWAP allows to continuously acquire 3D point cloud data of the environment while the robot is slowly moving forwards. This highly reduces the time needed to acquire a section of the underground mine, compared with the FARO scanner also mounted on the exploration vehicle. Such architectural scanners are usually meant to acquire the scene from a fixed position taking up to 10 minutes for scanning a single (but very dense) point cloud. An important development of this project is the mapping and registration tool mapit, which facilitates the registration and manipulation of point cloud and map data. We will outline the idea of mapit in Section 4. In Section 5 we present a number of visualization tools that were developed with mapit. In Section 6 we conclude.

2. Exploring and mapping underground mines

The underlying idea in the UPNS4D[1] project is to deploy two kinds of vehicles in an underground mining facility. There is an exploration vehicle that periodically drives around in the underground mine when no regular work is taking place to initially record and then update a map. The second vehicle is a regular processing vehicle that is performing the daily work in the mine. It used the map that is periodically updated by the exploration vehicle. While the exploration vehicle needs to have more sophisticated sensory equipment for recording the map, for the processing vehicle, a stripped-down equipment suffices, since it only needs to localize within a given map.

With our project partners, we developed the exploration robot shown in **Figure 1a**. It is a skid-steered tracked robot based on a mini excavator platform. It carries the modular sensor platform shown in **Figure 1b**. The robot can drive up to 3 ms^{-1} and is controlled via the ROS [17] Movebase. For navigation, collision avoidance and terrain classification, two Velodyne VLP-16 Puck LiDARs are mounted at the front. They acquire environment information with 16 scan lines with an opening angle of 30° and with 20 Hz. They are mounted on a 20° slope to be able to get information in the close vicinity of the robot. With a horizontal opening angle of 360°, they can acquire 3D data from the front and the sides of the robot. For safety reasons, additional 2D laser range finders have been mounted at two corners of the sensor platform which are also used for collision avoidance.

(a) (b)

Figure 1.
Exploration robot developed for mapping underground mining sites. (a) Exploration vehicle and (b) Sensor setup of the exploration robot.

	SWAP platform	FARO Focus^{3D} X 130
Measuring range	0.1–100 m	0.6–130
Horizontal resolution	2°	0.035°
Vertical resolution	0.4°	0.07°
Sphere coverage	80.27%	83.33%
Scan time	0.3–30 s	1–30 min

Table 1.
Comparison between SWAP and FARO Focus^{3D} X 130.

The mapping operation is not run autonomously in the mine environment for now. At the front of the robot, an Allied Vision GT6600C high-resolution camera with a wide-angle lens is mounted. The camera can be used for teleoperation.

As an additional safety feature, we mounted a FLIR A315 thermal camera at the front of the robot in order to be able to detect persons even when not sufficient light is available.

For mapping the mine, the platform is equipped with a rotating 3D LiDAR system, the SWAP platform, which we will describe in detail in the next section. For reference, we mounted a FARO Focus^{3D} X 130 LiDAR, which can be used in a stop-and-go fashion. Scanning times of the Focus LiDAR lie between 1 and 30 min. To remotely operate the LiDAR, we developed a ROS driver based on the FARO SDK.

As part of our project contribution, we developed a rotating sensor platform for the swift acquisition of dense point clouds as reported in [16]. The main goal was to find a compromise between acquiring accurate and dense point clouds which usually takes much time and having available data for online use in a robotic system for tasks such as localization which has to be updated more frequently. For instance, taking the FARO LiDAR with an angular resolution of 0.0035°, very dense and accurate point clouds can be recorded. However, the robot needs to stand still, and the scanning time of a single scan can take up to 30 min. **Table 1** shows a comparison of the two scanners.

3. The 3D LiDAR system SWAP

In this section, following previous work in [16], we present the 3D LiDAR platform SWAP which was developed during the mine mapping project. The SWAP platform consists of a Velodyne VLP-16 PUCK LiDAR and a Hokuyo UTM-30LX-EW range scanner which are both mounted opposite to each other on a disk which rotates both scanning devices around the centre of the disk. The disk and the upper part of the scanner are driven by a motor which is equipped with absolute encoders. Both scanners transfer their data via Ethernet which is connected by a slip ring which connects the revolving part to the rest of the scanning device. The combination of motor and gear head provides us with 3 Nm of torque and allows for a maximum rotation speed of 2.6 Hz. However, a reasonable azimuth resolution can only be achieved with a scanning speed of up to 1.67 Hz, while the full-sphere point clouds are then captured with a half revolution which equals 3.34 Hz for this. We deploy a 14 bit industrial grade absolute SSI encoder which is mounted on the drive shaft. The resolution provides a maximum error of 1.32′ or 0.022°. In a distance of 10, this corresponds to 3.8. The second part of the platform is the rotating sensor mount. It houses a gigabit Ethernet switch, the interface box of the Velodyne VLP-16 PUCK and the Hokuyo UTM-30LX-EW, the power distribution for the sensors and several mounting rails for different sensors. The raw data of the deployed Velodyne VLP-16 PUCK and the attached Hokuyo UTM-30LX-EW are registered making use of the SSI absolute encoder. Besides the absolute encoders, there is another incremental encoder attached to the motor shaft. Then, based on the readings of the absolute encoder, the raw data is collected and integrated into a point cloud for the device. This is done with a best-effort time-stamping on the data and where one UDP package of the Velodyne VLP-16 PUCK is transformed altogether. The time difference between the laser readings within one UDP package is about 1.33 ms. For the rectification of the Hokuyo UTM-30LX-EW measurements, the recording time for one sweep is taken into account. As a final ingredient, our SWAP platform is equipped with an IMU (μIMU from NG[1]) for providing the orientation of the platform w.r.t. the ground. **Figure 2** shows a CAD drawing as well as a photo of the device.

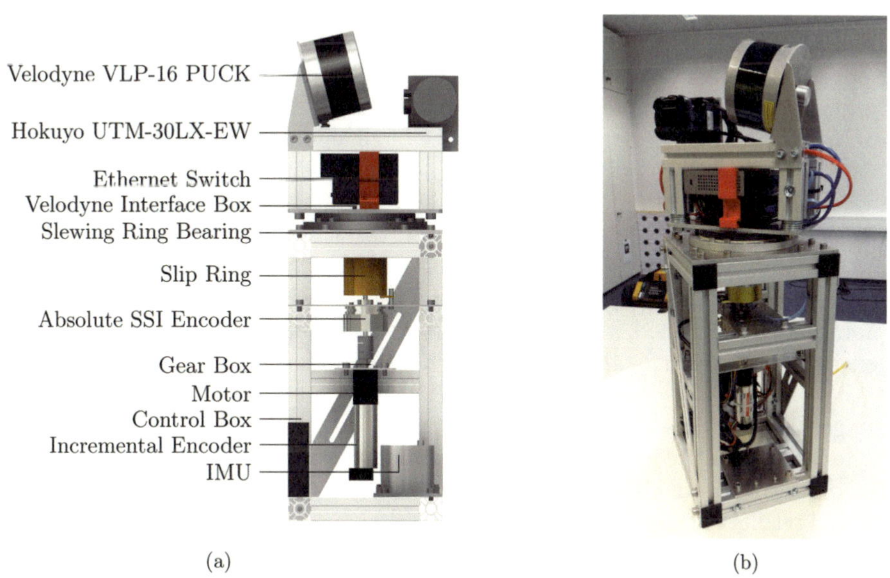

Figure 2.
The components and a photo of our rotating sensor platform. (a) Components of the platform and (b) Photo of the platform

[1] http://www.northropgrumman.litef.com/en/products-services/industrial-applications/product-overview/mems-imu/.

The current design was chosen keeping in mind the lessons learnt from its predecessor device which was based on a Velodyne HDL-64. That device was tilted along its vertical axis. It was presented in [18]. While the device is very suitable for acquiring 3D dense point clouds (in [7] we used this device for mapping large-scale motorway tunnels), it has some drawbacks when it comes to the distribution of the range measurement in the point cloud. With a tilting scanner, the point clouds are particularly dense at the turning points of the device and less dense in between.

For the SWAP design, we therefore changed the setup from a tilting 3D LiDAR to a rotating device. With the additional tilt angle in the mounting position of the VLP-16, we achieve an even distribution of points in the scanning range of the device. The additional Hokuyo rangefinder was mounted in order to acquire sensor data in a close range around the robot, as the measurement range of the VLP-16 starts at about 0.9 m.

Analyzing and optimizing the homogeneity of a scan are investigated in [19]. For our target scenarios, the sensor platform yields an optimized compromise between the scan acquisition speed and the point cloud density. By adjusting the rotation frequency, the map resolution and the time needed to record the map can also be balanced.

4. The map registration system mapit

There exist a number of toolkits that help with registering point clouds and process 3D data. Many of them are professional software products, often also provided from the manufacturer of a 3D LiDAR system. Also a number of Open Source projects exist, for instance, the 3D Toolkit (3DTK) [20, 21]. The 3DTK provides a number of state-of-the-art 6D SLAM algorithm for registering point cloud data as well as a large number of additional shape detection and visualization tools.

In contrast to this, mapit focuses more strongly on the registration workflow and the post-processing of map data, but it is not restricted only to point cloud data. In mapit, additional sensor cues could be associated with the 3D data. The key idea of mapit is to store the raw data and keep track of all algorithm steps over time.

4.1 Overview

With mapit, we developed a 3D mapping framework[2] for managing and post-processing a wide range of sensor data, especially the point clouds from the exploration vehicle. The software is divided into components and is designed for extensibility. We describe the different fields below:

Management/administration

The sensor data are loaded and stored persistently in a database. The access is via defined interfaces. All changes to the data are stored individually. As a result, work steps and results are stored together consistently.

Algorithm processing

The algorithms for filtering the sensor data, the registration tree of the 3D point clouds, the creation of the 3D map and the further processing of the map are defined and developed with mapit.

[2] https://github.com/MASKOR/mapit.

DOI: http://dx.doi.org/10.5772/intechopen.85859

Connection

A network interface has been developed that allows transparent access between local and remote mapit instances. A connection to external software (e.g. CloudCompare[3]) has been implemented to work efficiently through plugins.

While ROS is used to programme mobile robots and can save all sensor data from a robot test drive, the framework mapit was developed to manage and save the post-processing of all sensor data. There are three basic principles similar to a version control system that have been implemented in mapit:

1. Data and the executed algorithms are stored together: At any time, the origin of a map can be traced—the basic sensor data, the used algorithms and parameter for the post-processing.

2. Every access to the data must be done by mapit: This concept is like a version control system. The development and history of the post-processing are logged.

3. Algorithms are deterministic (if possible): Data can be deleted and restored at a later point.

All map data and algorithms can be stored like in a directory system. **Figure 3a** shows the structure of a management process, i.e. to develop a map. The repository corresponds to a project. Each project arranges its data in trees and entities, i.e. the point clouds. Each workspace consists of the post-processing algorithm workflow, i.e. to develop a map from registered point clouds.

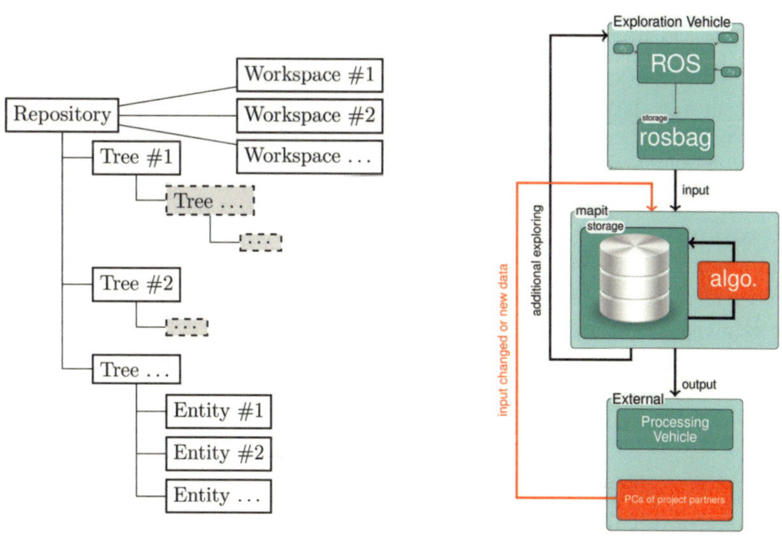

(a) (b)

Figure 3.
mapit concept and workflow. (a) Visualization of the mapit concept and (b) Workflow in mapit with the exploration vehicle and the processing vehicle involved.

[3] https://github.com/MASKOR/cc_qMapit_IO_plugin.

4.2 Workflow with mapit

This subsection describes the workflow with mapit to develop a 3D map from a test drive of the exploration vehicle. Mapit includes various requirements to support the work of 3D map creation. Firstly, it must have open and flexible interfaces to import and export sensor data and maps, as well as being easily adaptable to new requirements. Secondly, the data processing must be simple, reproducible and, in particular, traceable. Thirdly, it must have open and flexible interfaces to import and export sensor data and maps, as well as being easily adaptable to new requirements.

Figure 3b represents a mapit workflow from the exploration vehicle and the process vehicle. The figure has the following steps:

Exploratory trip

During the exploratory trip, every 10 full-sphere point clouds is recorded. These point clouds are stored together with the odometry data via the Robot Operating System (ROS) software with rosbag. After the exploratory trip, this data will be integrated into the mapit system.

Registration and mapping

In mapit, these point clouds are aligned to a consistent map using various registration algorithms (operators) and multiple passes. These aligned point clouds are then converted to a 3D occupancy map.

3D to 2D map conversion/transformation

The exploration vehicle and the production vehicle are located via 2D scanners, which measure the surface of the mine in one section. For these two vehicles, cuts are made in the mine's 3D map, and a 2D occupancy map is created.

Subsequent exploration

During the subsequent exploration, the exploration vehicle locates itself in this created 2D map or drives outside of it. New full-sphere point clouds of known and unknown areas are recorded and integrated into the mapit system after the trip to the previous one.

Map extension

Then, similar to the point *registration and mapping*, new point clouds are registered to the already aligned point clouds and the new extended map is then created with all data from previous reconnaissance trips and the new data from the additional reconnaissance.

4.3 3D mine mapping with mapit

To compute a consistent map based on the data collected by the mobile platform, we use spherical point clouds and minimize the error in merging them by integrating the robot's movements.

In order to integrate all sensor data to one global map, the algorithms in mapit do not operate on the data but only on the transformations on this data. This is why registration algorithms, for example, can be run on different resolutions or on different data without the data being modified: the results just change according to the modified transformations.

4.3.1 Pairwise registration

The point clouds recorded during the exploratory trip are first registered in pairs. For this, the iterative closest point (ICP) method [22] is used, which is an iterative minimum method. There are always two consecutive point clouds compared. It searches, from each point of the one point cloud, the next point in the other point cloud and minimizes this between all pairs of points minimized. The initials of the algorithm are the odometry provided by the AMT. Therefore, a general implementation has been created so that different algorithms, for example, can be easily integrated from the Point Cloud Library (PCL) [23]. To create the maps, the algorithm iterative closest point (ICP) has been used, which forms pairs of points between two point clouds and minimizes the distance between these pairs. The initial values for these algorithms are primarily the odometry provided by the RWTH Aachen University Institute for Advanced Mining Technologies (AMT). This results in a good orientation of the point clouds but typically remains a small

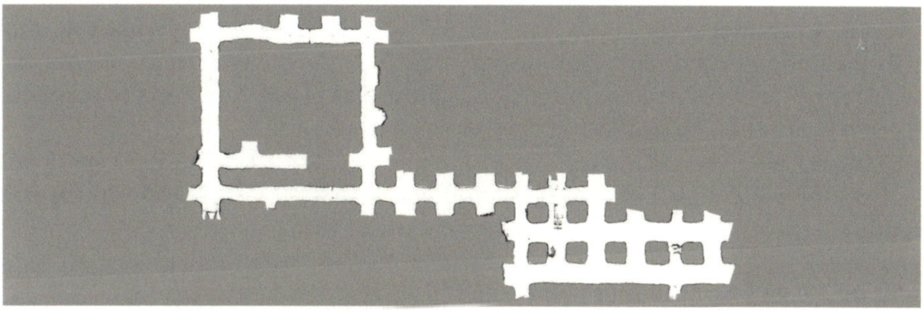

(a)

(b)

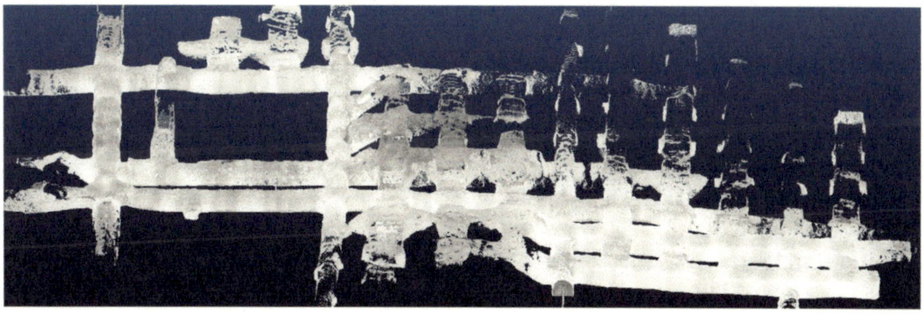

(c)

Figure 4.
Parts of the anhydride mine in Krölpa, Germany. (a) Occupancy grid, (b) Octree and (c) Point Cloud.

error. This becomes visible after long scan series (e.g. in the map of maxit, Krölpa, of approximately 800 m range).

4.3.2 Global registration

To minimize the errors of pairwise registration of many point clouds, all point clouds are registered globally. Again, the algorithms are easily interchangeable. Mostly currently the GraphSLAM [24] algorithm is used after Lu and Milios [25]. It creates a graph of the connections between all point clouds with overlaps and minimizes the alignment errors of all connections simultaneously.

4.3.3 Conversion to 3D map

As a representation of the 3D map, an OctoMap [26] is used. This offers the advantage of being able to query the information in various resolutions and to map the distinctions that are important for navigation between free, occupied and unknown cells.

Figure 4 represents the result from the 3D mapping process from an exploratory trip in the underground mine from maxit in Krölpa, Germany. The mobile robot scans every 10 m a full-sphere point cloud. The point cloud data and the odometry data are saved into a rosbag file and are processed with the mapit workflow. **Figure 4a** shows a 2D occupancy grid of the mapped part of the mine. **Figure 4c** visualizes the point clouds themselves and **Figure 4b** a 3D OctoMap from the point clouds.

5. Visualization

ROS includes a 3D visualization environment package RViz that fuse sensor data, robot model and other 3D data like point clouds into a combined view. RViz uses several external libraries like the Ogre3D graphics library for the 3D visualization. The 3D points from the point clouds can be visualized as small surface elements (surfel) or boxes. A surface mesh can be developed with surface reconstruction algorithms. This very expensive step is mostly avoided and is approximated by choosing the size of the surfels or boxes.

For the post-processing of the robot data, a graphical user interface (GUI) for mapit has been developed. The user can define a new project and can select all considered data over the client-server connection. Mapit has implemented algorithms to register point clouds, space decomposition algorithm for efficient computation and rendering algorithm for the visualization. All user-selected algorithms can be arranged and saved in a workflow and can be visualized and edited in the GUI as a node graph. **Figure 5** shows the mapit GUI and the node graph to calculate and render 13 point clouds from the ground floor of the main building of the Aachen University of Applied Sciences.

5.1 Visualization from sensor and mapping data

For almost all sensors attached to the exploration vehicle, ROS connections are provided for visualization. The calibration software of the SWAP platform allows a visualization of the point clouds recorded over time in near real-time. The raw data of the FARO laser scanner are also automatically processed on-board on-site and made available within ROS. This process takes a few seconds to a few minutes, depending on the volume of data. **Figure 6a** represents a combined visualization

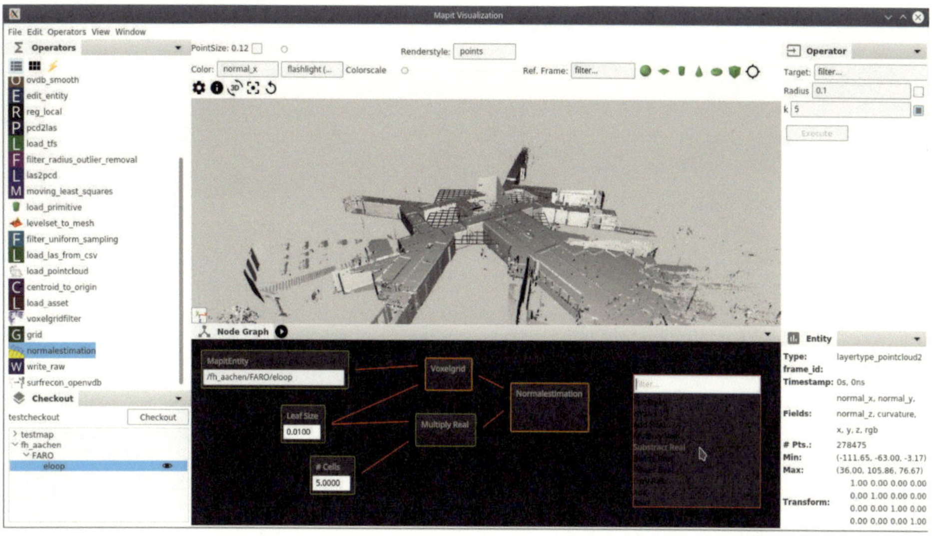

Figure 5.
Visualization from the mapit GUI to render 13 registered point clouds.

from the stop-and-go FARO scan point cloud data in gray values and the live sensor data over 2 seconds from the SWAP platform in color values. **Figure 6b** shows in a top-down view the real-time data from one Velodyne Puck scanner with 20 Hz in blue and red color values.

The exploration vehicle can scan automatically with the automation adapter of the FARO Focus laser scanner. A main disadvantage of this automation adapter is that all drivers for the FARO scanner support only the Windows operating system. To overcome this problem, several Windows applications are developed and can be selected via ROS. Firstly, the scanning application sends user-selected scan parameter over ROS to the scanner and starts the scanning process. The scanned data are stored in the customer format on the hard disk. The second application converts the customer format of a point cloud into the free PCD format of the Point Cloud Library (PCL) and stores the new format again. Last but not least, the PCD data files are loaded and are visualized with RViz in ROS. As an option, the PCD data can be filtered.

The huge amount of point cloud data can be decomposed with an octree space partitioning algorithm (see **Figure 7**). This data structure is suitable for local collision detection, downsampling the huge data amount and representing the data in different resolutions. Especially, the normal estimation uses a neighboring search to every point of the point cloud. This step and the viewer-dependent resolution can be calculated more efficiently using the octree data structure.

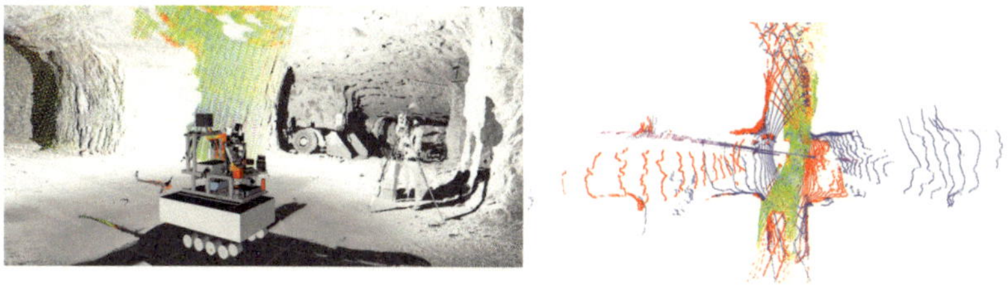

(a) (b)

Figure 6.
Live sensor data from the SWAP platform of the exploration vehicle at the maxit mine, Krölpa, Germany. (a) Grey values: Stop-and-go data from FARO Focus3D X 130. Color values: Live data from SWAP platform over 2s and (b) Top down perspective from Velodyne VLP-16 puck data, 20Hz in blue and reds.

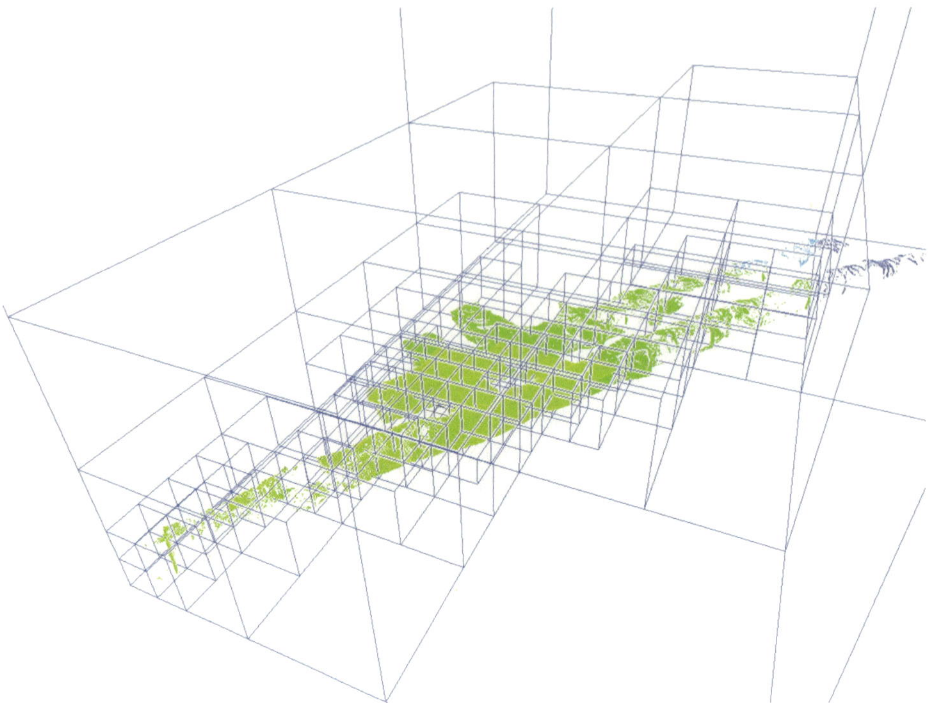

Figure 7.
Spatial decomposition of the point cloud with an octree data structure. The points are organized in a hierarchical fashion where the resolution of space is increased only when the space actually contains data points.

An interface from mapit to ROS allows a generated map to be returned to the context of the exploration vehicle. This allows current sensor data to be visualized in a corresponding map section with appropriate resolution during a re-exploration.

5.2 Virtual reality integration with CuteVR

The CuteVR library was developed as a bridge between the virtual reality hardware drivers and the end-user software with the aim of a consistent interface. CuteVR is highly modular based on the cross-platform application framework Qt and can be extended due to its class structures with relatively little effort. An event system and differentiated error handling also make it easier to handle highly dynamic VR scenes.

It forms the basis for a VR plugin for RViz, which allows the viewing of all sensor data in VR. This allows the user to navigate in the virtual reality world next to the exploration vehicle (see **Figure 8**). CuteVR unifies the interfaces to VR devices and can be expanded for future VR hardware.

Based on CuteVR, the ROS package vr_tools was developed, which integrates VR hardware and VR concepts in ROS. The core component is the head-mounted display (HMD) plugin for RViz, with which one or more users can look around and move around in a RViz scene. This allows intuitive and true-to-scale viewing of 3D sensor data in the virtual space.

Since RViz itself does not provide structures for the spatial distribution of large amounts of data such as octrees, it was decided instead to filter the data stream to RViz and adapt it on the fly.

Furthermore, the states of VR input devices are made available in ROS and thus can interact with other programme components. With additional VR setups,

Figure 8.
Virtual reality visualization with CuteVR using the HTC Vive VR headset.

multiple users can view the same scenes simultaneously from different perspectives via Multi-View.

6. Discussion

In this chapter, we presented a system for continuous mapping and exploration of underground sites. Most of this work has been developed as part of the project "Underground 4D+ Positioning, Navigation and Mapping System for Highly Selective, Efficient and Highly-secure Exploitation of Important Resources" (UPNS4D+) which was funded by the German Federal Ministry of Education and Research within the programme of "R4–Innovative Technologies for Resource Efficiency – Research for the Provision of Raw Materials of Strategic Economic Importance".

We first reported on the hardware platform that was built to acquire comparably densely populated 3D point clouds of the (underground) environment using a rotating LiDAR device. Afterwards, we reported on the framework mapit which is used to track and execute post-processing operations on the data acquired by the robot. Namely, it allows for registering a (large) set of individual maps to one global map. The important aspect is that mapit does not store the resulting map and discards the original data, but instead it keeps track of the operations that were performed on the data and logs these operations. This allows for reapplying all post-processing in case an algorithm has been improved or a misalignment in the calibration of the sensor setup has been detected. Finally, we presented the options for visualizing the resulting maps in different contexts.

The system described in this chapter provides diverse support for (first) responders in search and rescue applications. For one, the resulting maps can be used to conduct further missions with rescue robots. Also, analysis tools can be run on the maps. For example, the mapit framework supports running algorithms to compare maps from different points in time to see which changes have occurred. The versatile visualization capabilities allow for planning rescue missions and training first responders before sending them into the field.

Acknowledgements

This research was funded by the German Federal Ministry of Education and Research within the programme of "R4–Innovative Technologies for Resource

Efficiency—Research for the Provision of Raw Materials of Strategic Economic Importance" and is part of the project UPNS4D+ with FKZ 033R126C. Special thanks go to Tobias Hartmann from the Advanced Mining Technology Institute of RWTH Aachen University and Prof. Paul Burgwinkel and Jörn Lachmann from the company Fritz Rensmann GmbH & Co. with whom we designed and built up the exploration vehicle.

Further, we want to thank the other consortium partners for their valuable contributions: indurad GmbH; MILAN Geoservice GmbH; Institute for Advanced Mining Technology at RWTH Aachen University; XGraphic Ingenieurgesellschaft mbH, Aachen; Technische Universität Bergakademie Freiberg; Fritz Rensmann, Maschinenfabrik, Diesellokomotiven, Getriebe GmbH & Co. KG; and GHH Fahrzeuge GmbH.

Author details

Alexander Ferrein, Ingrid Scholl, Tobias Neumann, Kai Krückel and Stefan Schiffer*
Mobile Autonomous Systems and Cognitive Robotics Institute (MASCOR),
FH Aachen University of Applied Sciences, Aachen, Germany

*Address all correspondence to: s.schiffer@fh-aachen.de

References

[1] Kruijff G-JM, Pirri F, Gianni M, Papadakis P, Pizzoli M, Sinha A, et al. Rescue robots at earthquake-hit Mirandola, Italy: A field report. In: 2012 IEEE international symposium on Safety, Security, and Rescue Robotics (SSRR), IEEE. 2012. pp. 1-8

[2] Kruijff G-JM, Kruijff-Korbayová I, Keshavdas S, Larochelle B, Janček M, Colas F, et al. Designing, developing, and deploying systems to support human–robot teams in disaster response. Advanced Robotics. 2014; **28**(23):1547-1570

[3] Mascarich F, Khattak S, Papachristos C, Alexis K. A multi-modal mapping unit for autonomous exploration and mapping of underground tunnels. In: 2018 IEEE Aerospace Conference. March 2018. pp. 1-7

[4] Erdelj M, Natalizio E, Chowdhury KR, Akyildiz IF. Help from the sky: Leveraging UAVS for disaster management. IEEE Pervasive Computing. 2017;(1):24-32

[5] Pellenz J, Lang D, Neuhaus F, Paulus D. Real-time 3D mapping of rough terrain: A field report from disaster city. In: IEEE International Workshop on Safety Security and Rescue Robotics (SSRR); IEEE. 2010. pp. 1-6

[6] Ohno K, Tadokoro S, Nagatani K, Koyanagi E, Yoshida T. Trials of 3-d map construction using the tele-operated tracked vehicle kenaf at disaster city. In: 2010 IEEE International Conference on Robotics and Automation (ICRA). IEEE; 2010. pp. 2864-2870

[7] Leingartner M, Maurer J, Ferrein A, Steinbauer G. Evaluation of sensors and mapping approaches for disasters in tunnels. Journal of Field Robotics. 2016; **33**(8):1037-1057

[8] Kohlbrecher S, Meyer J, Graber T, Petersen K, Klingauf U, von Stryk O. Hector open source modules for autonomous mapping and navigation with rescue robots. In: Behnke S, Veloso M, Visser A, Xiong R, editors. RoboCup 2013: Robot World Cup XVII, Berlin, Heidelberg, Springer Berlin Heidelberg. 2014. pp. 624-631

[9] Dubé R, Gawel A, Cadena C, Siegwart R, Freda L, Gianni M. 3D localization, mapping and path planning for search and rescue operations. In: 2016 IEEE International Symposium on Safety, Security, and Rescue Robotics (SSRR). Oct 2016. pp. 272-273

[10] Balta H, Velagic J, Bosschaerts W, De Cubber G, Siciliano B. Fast iterative 3D mapping for large-scale outdoor environments with local minima escape mechanism. In: 12th IFAC Symposium on Robot Control SYROCO 2018. IFAC-PapersOnLine, Vol. 51(22). 2018. pp. 298-305

[11] Murphy RR, Tadokoro S, Kleiner A. Disaster robotics. In: Springer Handbook of Robotics. Springer; 2016. pp. 1577-1604

[12] Zhang Z, Nejat G, Guo H, Huang P. A novel 3D sensory system for robot-assisted mapping of cluttered urban search and rescue environments. Intelligent Service Robotics. 2011;**4**(2): 119-134

[13] Larochelle B, Kruijff G-JM. Multi-view operator control unit to improve situation awareness in USAR missions. In: 2012 IEEE, RO-MAN; IEEE. 2012. pp. 1103-1108

[14] Buttgereit DA, Hartmann T, Schade S, Nienhaus K. Auf dem Weg zu nachhaltigen Abbauprozessen mit UPNS 4D+547-8. In: 20. Kolloquium Bohr- und Sprengtechnik: Institut für

Bergbau, Technische Universität Clausthal; Clausthal-Zellerfeld (Germany); 18. und 19. Januar 2017: Tagungsband/Herausgeber: Univ.-Prof. Dr.-Ing. Oliver Langefeld, Univ.-Prof. Dr.-Ing. habil. Hossein Tudeshki; Clausthal-Zellerfeld, 18 Jan 2017–19 Jan 2017. 1. Auflage ed. 2017. pp. 131-140. Papierflieger

[15] Donner R, Rabel M, Scholl I, Ferrein A, Donner M, Geier A, et al. The extraction of relevant features from 3D point clouds of a mobile multi-sensor system in an underground mine setting. In: Tagungsband Geomonitoring 2019; Institutionelles Repositorium der Leibniz Universität, Hannover. 2019. pp. 91-110 (in German)

[16] Neumann T, Dülberg E, Schiffer S, Ferrein A. A rotating platform for swift acquisition of dense 3D point clouds. In: ICIRA (1), volume 9834 of Lecture Notes in Computer Science. Springer; 2016. pp. 257-268

[17] Quigley M, Conley K, Gerkey BP, Faust J, Foote T, Leibs J, et al. ROS: An open-source robot operating system. In: ICRA Workshop on Open Source Software. 2009

[18] Neumann T, Ferrein A, Kallweit S, Scholl I. Towards a mobile mapping robot for underground mines. In: Proceedings of the 7th IEEE Robotics and Mechatronics Conference (RobMech-14). 2014

[19] Mandow A, Morales J, Gomez-Ruiz JA, García-Cerezo AJ. Optimizing scan homogeneity for building full-3D lidars based on rotating a multi-beam velodyne range-finder. In: 2018 IEEE/RSJ International Conference on Intelligent Robots and Systems (IROS). Oct 2018. pp. 4788-4793

[20] Nüchter A, Lingemann K. 3DTK—The 3D Toolkit. 2011. Available from: http://slam6d.sourceforge.net/doc/slam6ddoc.html

[21] Nüchter A. 3D robotic mapping. In: volume 52 of Springer Tracts in Advanced Robotics. Springer; 2008

[22] Besl P, McKay ND. A method for registration of 3-D shapes. IEEE Transactions on Pattern Analysis & Machine Intelligence. 1992;**14**(2): 239-256

[23] Rusu RB, Cousins S. 3D is here: Point Cloud Library (PCL). In: 2011 IEEE International Conference on Robotics and Automation. 2011. pp. 1-4

[24] Thrun S, Montemerlo M. The graph SLAM algorithm with applications to large-scale mapping of urban structures. The International Journal of Robotics Research. 2006;**25**(5–6):403-429

[25] Lu F, Milios E. Robot pose estimation in unknown environments by matching 2d range scans. Journal of Intelligent and Robotic Systems. 1997; **18**(3):249-275

[26] Hornung A, Wurm KM, Bennewitz M, Stachniss C, Burgard W. OctoMap: An efficient probabilistic 3D mapping framework based on octrees. Autonomous Robots. 2013;**34**(3): 189-206

Chapter 5

Advanced UAVs Nonlinear Control Systems and Applications

Abdulkader Joukhadar, Mohammad Alchehabi and Adnan Jejeh

Abstract

Recent development of different control systems for UAVs has caught the attention of academic and industry, due to the wide range of their applications such as in surveillance, delivery, work assistant, and photography. In addition, arms, grippers, or tethers could be installed to UAVs so that they can assist in constructing, transporting, and carrying payloads. In this book chapter, the control laws of the attitude and position of a quadcopter UAV have been derived basically utilizing three methods including backstepping, sliding mode control, and feedback linearization incorporated with LQI optimal controller. The main contribution of this book chapter would be concluded in the strategy of deriving the control laws of the translational positions of a quadcopter UAV. The control laws for trajectory tracking using the proposed strategies have been validated by simulation using MATLAB®/Simulink and experimental results obtained from a quadcopter test bench. Simulation results show a comparison between the performances of each of the proposed techniques depending on the nonlinear model of the quadcopter system under investigation; the trajectory tracking has been achieved properly for different types of trajectories, i.e., spiral trajectory, in the presence of unknown disturbances. Moreover, the practical results coincided with the results of the simulation results.

Keywords: UAVs, nonlinear control, quadcopter, gesture-based vision control, spherical blimp UAV

1. Introduction

Unmanned aerial vehicle (UAV) research has attracted tremendous attention during the last decade. This interest is mainly given due to the low cost of this type of vehicles and its large application range in diverse areas such as surveillance, delivery, maintenance, inspection, transportation, work assistant, and aerial photography. For instance, UAVs could be provided with cameras so as to observe nature and wildlife. In addition, arms, grippers, or tethers might be installed to UAVs, for which UAVs can assist in construction, transportation, and carrying payloads. Different types of UAVs are considered as complex systems since their dynamic models are nonlinear, dynamically coupled, and the difficulty to establish a very accurate mathematical model. The design of UAV control systems has attracted many researchers worldwide, and many control techniques have been proposed for the aim of accomplishing a 6-DoFs dynamic and trajectory tracking control of UAVs. This chapter focuses on advanced nonlinear control approaches in order to enhance the dynamic performance of both dynamic and trajectory tracking control of UAVs. Nonlinear control theories have been developed among other control strategies due to their capacity to deal with the nonlinearity and the coupling components of the UAV state variables.

Quadcopters are one of the very common UAV platforms; in fact, the literature related to control design of quadcopters is extensive, and this type of UAVs is underactuated, nonlinear, and strongly coupled, which is hard to cope properly with conventional control methods. On the other hand, they have many advantages over conventional helicopters, which may be concluded as follows: capability of vertical take-off and landing (VTOL), hovering and maneuverability, and low power consumption, since it has four small-scale propellers for thrust and orientation.

In the area of quadcopter literature, there is a variety of applications as aerial manipulation [1, 2], quadcopter pendulum [3], navigation and localization [4, 5], obstacle avoidance [6], altitude control [7], and cooperative and formation control [8, 9]. Moreover, several control schemes have been proposed including adaptive control [10–13], fuzzy control [14], neural network control [15], linear parameter varying (LPV) control [16], predictive control [17, 18], nonlinear control methods [19–23], and sliding mode control [24, 25]. In [4], researchers propose localization, navigation, and mapping methods based on the characteristic map; feature map is selected to localize and navigate the UAV under investigation, while drawing up navigation strategy and avoidance strategy. In [5], PID controllers for the attitude, altitude, and position of a quadrotor are designed, and an outdoor experiment is conducted based on GPS to verify the performance, and desired trajectory's waypoints are determined using Mission Planar software. The application of ultrasonic sensor is used to detect barriers during the flight, so that the position of the quadrotor is adjusted depending on the signal of the ultrasonic sensor in order to avoid collision [6]. Cascaded PID controller with the usage of laser range finder combined with accelerometer in order to determine the height of the vehicle has been presented in [7]; the proposed system is compared with the performance of the system using GPS combined with pressure gauge. However, the results of the proposed system exhibit better performance especially in the range of low altitude. Centralized formation flight control of a leader/follower structure of three quadcopters is proposed in [8] using LQR-PI, the trajectory of the leader defines the desired trajectory for the followers, and a pole placement controller is used for the leader and LQR-PI controllers for the followers. In case of communication loss between leader and any of the followers, the other follower quadcopter provides the leader's states to the affected follower quadcopter in order to keep the formation intact. Whereas a multiagent consensus control incorporated with collision avoidance using model predictive control is presented in [9], the term of achieving formation and the term of repulsive potential are set in the index function to realize the formation control considering collision avoidance. The experiment is carried out using three quadrotor UAVs.

By looking to the quadcopter control systems, dynamic inversion and linear neural-network-based adaptive attitude control of a quadrotor UAV is introduced in [10]. Based on the time-scale separation principle, an attitude dynamic inverse controller and a trajectory dynamic inverse controller are deduced, respectively; the inverse error dynamics is regulated using PD controller, and a sigma-pi neural network is introduced to eliminate the inverse error adaptively to improve the robustness of the controller. Authors of [11] propose a compound adaptive backstepping and sliding mode control subject to unknown external disturbances and parametric uncertainties. A comparison study for the proposed method with and without adaptive control is investigated. An adaptive controller based on backstepping technique is employed for the trajectory tracking of quadrotor incorporating a fuzzy monitoring strategy to compensate the undesired dynamic error caused by lumped disturbances and total thrust input saturation [12]. Reference [13] introduces adaptive sliding mode controller for distributed control systems with mismatched uncertainty that exists in communication channels. A linear sliding surface is adopted to guarantee asymptotic stability of each subsystem, and an adaptive scheme that can update the unknown upper bound of uncertainty is applied. The distributed controller is constructed based on the information from the adaptive scheme and neighboring subsystems, such that each subsystem can keep stable and have good performance.

On the other hand, a tracking control system for the quadrotor UAV based on Takagi-Sugeno (T-S) fuzzy control has been presented in [14]. At first, T-S fuzzy error model has been presented as three independent subsystems for altitude, attitude, and position. Then, T-S fuzzy feedback controller design procedure is applied for altitude, attitude, and position subsystems of the quadcopter. LMI algorithm has been utilized in order to calculate the controller's gains. In [15], a sliding mode underactuated control (SMUC) is designed for the quadrotor UAV model with small uncertainty. In order to enhance the tracking response of the quadrotor UAV, recurrent-neural-network-based sliding-mode underactuated control (RNN-SMUC) with online recurrent neural network modeling and compensation of dynamical uncertainty is designed, and the RNN performs as an approximator. Finally, the combination of SMUC and RNN-SMUC with a transition as so-called hybrid neural-network-based sliding-mode underactuated control (HNN-SMUC) is developed. This development has the advantages of SMUC and RNN-SMUC; e.g., a better transient response of SMUC and an improved tracking performance of RNN-SMUC are accomplished. Furthermore, researchers of [16] compare between LPV controllers and LTI H_∞ controllers with S/KS loop shaping to test the performance of a quadrotor while tracking fast trajectories and aggressive maneuvers. Reference [17] combines nonlinear model predictive control (NMPC) and PID controller for better stabilizing of quadrotor UAV under different noises and disturbance conditions; the proposed controller has been applied for the altitude and attitude control loops, whereas switching model predictive controllers for attitude, altitude, and translational motion are derived based on piecewise affine linearized dynamic model in [18], where the effects induced by wind gusts disturbances are considered as affine outputs. The experimental platform utilizes inertial measurement unit IMU, sonar and an optic-flow sensor to produce feedback to the system for indoor applications. Various flight cases including position hold and altitude set-point, trajectory tracking, hovering, and aggressive attitude control have been performed in order to justify the efficiency of the proposed control system.

Nonlinear control methods cover the majority of the applied approaches in the literature. For instance, [19] proposes nonlinear hybrid controller that utilizes the time response characteristics of the PID and the stability characteristics of the LQR; differential-flatness-based feedforward control is incorporated with the LQR to enhance the performance of the position system, whereas PID controllers are designed to control the attitude of the quadcopter. Authors of [20] utilize LQR, PID, and feedback linearization in order to design position-tracking model. The LQR controller is added to the feedback linearization model to optimize the control algorithm by determining a suitable cost function; the attitude of dynamic control was modeled so as to maintain desired quadcopter's position despite the presence of disturbances. The performance of tracking position is optimized by adding PID loop control for pitch, roll, and yaw movement, and a comparison between the performance of the two nonlinear control techniques, including backstepping and feedback linearization with LQR, has been performed in [21]. The control laws have been derived depending on the nonlinear model with no linearization, and experiments for the attitude have been performed. Whereas in [22], the performance of sliding mode techniques has been verified and *sat* function has been used in order to obtain a continuous control law instead of *sign* function [23]. This shows nonlinear control laws applied for optimal trajectory tracking depending on minimum snap theory, and differential flatness method is utilized to derive control laws that link between the system outputs and its inputs. Reference [24] focuses on sliding mode control of the quadcopter; the proposed approach consisted of a sliding mode observer with finite-time process, a hybridization of a PID conventional controller, and a continuous sliding-mode one. The main aim is to estimate the system's state vector based on the measured system's output states and to identify a certain type of the inherited system's disturbances simultaneously. It is also to track a desired time-varying trajectory in spite of the influence of external disturbances and uncertainties. Finally, fractional order sliding mode control is used to derive the attitude control laws of a quadcopter, where PD tracking controllers are used to control the position of the quadcopter in [25].

2. Nonlinear control approaches

Nonlinear control theory is the area of control theory that deals with nonlinear systems, time variant systems, or both. Different engineering applications motivate researchers to develop powerful nonlinear control methods, since a majority of these systems are considered to be nonlinear. The key reason behind the use of nonlinear control techniques is their capability to deal with the nonlinear characteristics of nonlinear systems such as underactuations, models uncertainty, and dynamic coupling. This chapter focuses on the following nonlinear control approaches:

1. Backstepping
2. Sliding mode control
3. Feedback linearization

2.1 Backstepping

It is a widely used nonlinear control technique, due to its significant inherited characteristics including: being a recursive controller approach, which depends on a proposed Lyapunov function for deriving the system control law; higher flexibility, to some extent, in avoiding key nonlinearity cancellation; and verifying the desired objective of stabilization and tracking [26, 27]. The procedure of deriving control laws depending on backstepping technique is concluded in, at first, determining the error function between the desired input and the system actual output, then outlining a Lyapunov function and determining virtual controls to make the derivative of the proposed Lyapunov function with a negative definite. Finally, these steps are repeated until obtaining the control law.

Consider the following system:

$$\begin{aligned} \dot{x}_1 &= x_2 \\ \dot{x}_2 &= f(x) + g(x)u \\ y &= x_2 \end{aligned} \tag{1}$$

Let us define the following error function as

$$e_1 = x_{1d} - x_1 \tag{2}$$

and Lyapunov function as

$$V = \frac{1}{2} e_1^2 \tag{3}$$

In order to obtain the derivative of the proposed Lyapunov function with a negative definite,

$$\dot{V} = -ke_1^2 + e_1(x_{1d} + k_1 e_1 - f(x) - g(x)u) \tag{4}$$

where k is a positive constant, so that the control law will be as follows:

$$u = \frac{1}{g(x)}(x_{1d} + ke_1 - f(x)) \tag{5}$$

Note: it is remarkable to mention that the parameters of a system would appear in the derived control law when using backstepping, so that an integral action is added to each virtual control during the procedure of deriving the control law, which is termed integral backstepping, and more details about backstepping method are described in [27].

2.2 Feedback linearization

Feedback linearization is also one of the major nonlinear design tools. It is used to cancel the nonlinear terms in a system's model; this cancellation resulting in a linear system allows designing and incorporating linear controllers for a nonlinear system with the feedback linearization laws. To introduce the procedure of this strategy, we first introduce the notions of *full-state linearization*, where the state equation is completely linearized, and *input-output linearization*, where the input-output map is linearized, while the state equation may be only partially linearized [26].

In this chapter, we will pay attention to input-output linearization method. To obtain the input-output feedback linearization law, we simply repeat the calculation of the derivative of the system output along the state variables. Let us consider the system in (1) as,

$$y = x_2, \dot{y} = \dot{x}_2 \tag{6}$$

The input-output linearization law would become:

$$u = \frac{1}{g(x)}(-f(x) + v) \tag{7}$$

2.3 Sliding mode control

Sliding mode control is considered one of the control tools of the variable structure systems (VSS), since it produces a discontinuous controller. It has the advantage of stabilizing and achieving robustness criteria against model uncertainty and disturbances. Sliding mode control theory depends on a sliding surface s, where the sliding mode controller constrains a system to it. The motion toward the sliding surface consists of a *reaching phase* during which trajectories starting off

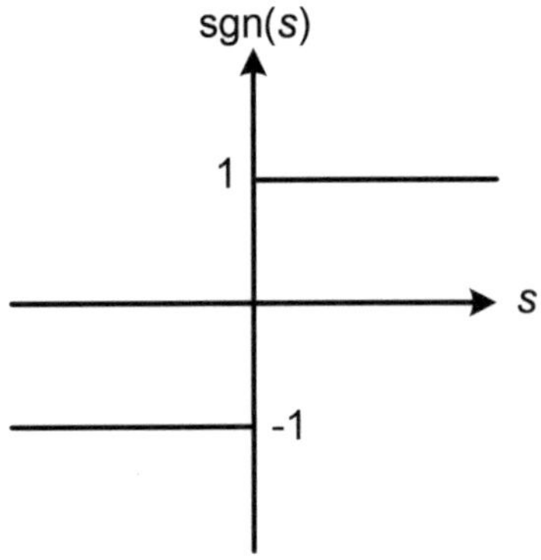

$$\operatorname{sgn}(s) = \begin{cases} 1 & s > 0 \\ 0 & s = 0 \\ -1 & s < 0 \end{cases}$$

Figure 1.
Illustration of sign function.

the surface $s = 0$ move toward it and reach it in finite time, followed by a *sliding phase* during which the motion is confined to the surface s [26, 28].

Equivalent control law is one of the sliding mode control strategies; it consists of two terms, the first is produced by equaling the derivative of sliding surface s to 0. The other term is called reaching law that has some common formulas such as [28]:

Constant rate reaching law, i.e., $\dot{s} = -K \operatorname{sgn}(s)$,

and constant plus proportional rate reaching law, i.e., $\dot{s} = -Qs - K \operatorname{sgn}(s)$.

where Q and K are positive constants and sign(s) is illustrated in **Figure 1**.

With $V = \frac{1}{2}s^2$ as a Lyapunov function candidate, hence the condition of the stability is $\dot{V}$ to be negative definite. In order to ensure that error e converges to zero, the sliding surface might be supposed as a function of the error as follows [26]:

$$s = c_0 e + c_1 \dot{e} + \ldots\ldots + c_{d-1} e^{\rho - 1(.)} + e^{\rho(.)} \tag{8}$$

where ρ is the relative degree.

The procedure for designing a sliding mode controller can be summarized by the following steps:

1. Designing the sliding surface s

2. Determining the derivative of the sliding surface $\dot{s}$

3. Equaling the derivative of sliding surface with the appropriate reaching law

4. Deriving the control law from the previous step

3. Quadcopter modeling

The dynamic model of the quadcopter is delivered in this section; the details of the model can be seen in the literature [29–31]. The state variables of the quadcopter are defined as, $X = \left[x, y, z, \dot{x}, \dot{y}, \dot{z}, \varphi, \theta, \psi, \dot{\varphi}, \dot{\theta}, \dot{\psi}\right]^T$ where $\zeta = [x, y, z]^T$ is the position described in the inertial coordinate frame B, $V = [\dot{x}, \dot{y}, \dot{z}]^T$ is the translational velocity, $\eta = [\varphi, \theta, \psi]^T$ are the roll-pitch-yaw angles describing the attitude of the quadcopter, and $\dot{\eta} = [\dot{\varphi}, \dot{\theta}, \dot{\psi}]^T$ are the Euler angle rates of the quadcopter described in the body-fixed frame A.

where ${}^{B}R_A$ is the transformation matrix

$$ {}^{B}R_A = \begin{bmatrix} c\psi c\theta & s\varphi.s\theta.c\psi - c\varphi s\psi & c\varphi.s\theta.c\psi + s\varphi.s\psi \\ s\psi c\theta & s\varphi.s\theta.s\psi + c\varphi c\psi & c\varphi.s\theta.s\psi - s\varphi.c\psi \\ -s\theta & s\varphi c\theta & c\varphi c\theta \end{bmatrix} \tag{9} $$

The equations of motion can be written as follows (10):

$$ \begin{aligned} \ddot{x} &= (c\varphi s\theta c\psi + s\varphi s\psi)\frac{U_1}{m} \\ \ddot{y} &= (c\varphi s\theta s\psi - s\varphi s\psi)\frac{U_1}{m} \\ \ddot{z} &= (c\varphi c\theta)\frac{U_1}{m} - g \\ \ddot{\varphi} &= \frac{I_{zz} - I_{yy}}{I_{xx}}\dot{\theta}\dot{\psi} + \frac{J_r}{I_{xx}}\Omega_r\dot{\theta} + \frac{1}{I_{xx}}U_2 \\ \ddot{\theta} &= \frac{I_{zz} - I_{xx}}{I_{yy}}\dot{\varphi}\dot{\psi} - \frac{J_r}{I_{yy}}\Omega_r\dot{\varphi} + \frac{1}{I_{yy}}U_3 \\ \ddot{\psi} &= \frac{I_{xx} - I_{yy}}{I_{zz}}\dot{\varphi}\dot{\theta} + \frac{1}{I_{zz}}U_4 \end{aligned} \tag{10} $$

where m is the mass of the quadcopter given in kilograms. With,

$$ \begin{aligned} U_1 &= f_1 + f_2 + f_3 + f_4 \\ U_2 &= l(f_4 - f_2) \\ U_3 &= l(f_3 - f_1) \\ U_4 &= T_1 - T_2 + T_3 - T_4 \end{aligned} \tag{11} $$

where $f_i = b\omega_i^2$ is the thrust force produced by propeller i with thrust coefficient b in N·s^2/m and ω_i is the angular speed of motor i.

$T_i = d\omega_i^2$ is the drag torque produced by propeller i in N·m with corresponding drag coefficient d in N^{2}s, l is the distance between center of the quadcopter and center of propeller in m, I is the inertia matrix, and I_{xx}, I_{yy}, and I_{zz} are moments of inertia about x, y, and z axes, respectively, in kg·m^2.

where J_r is the moment of inertia of the propeller and Ω_r is the sum of the four motors' angular speed. Based on the above derivation and discussion,

$$ U = \begin{bmatrix} U_1 \\ U_2 \\ U_3 \\ U_4 \end{bmatrix} = \begin{bmatrix} b & b & b & b \\ 0 & -bl & 0 & bl \\ -bl & 0 & bl & 0 \\ d & -d & d & -d \end{bmatrix} \begin{bmatrix} \omega_1^2 \\ \omega_2^2 \\ \omega_3^2 \\ \omega_4^2 \end{bmatrix} \tag{12} $$

4. Quadcopter control

The control scheme of the quadcopter can be represented as in **Figure 2**, it consists of two loops: the attitude control loop and the inner loop, which produces the control commands for the quadcopter to move. Moreover, the position control loop and the outer loop produce the references for the inner loop.

In this section, the control laws of the quadcopter will be derived using the aforementioned nonlinear control methods.

4.1 Quadcopter control using integral backstepping

Control laws of the attitude and position of the quadcopter are derived using integral backstepping approach.

4.1.1 Altitude control

We will start deriving the control law of the attitude by defining the altitude error and the Lyapunov function as follows:

$$e_1 = z_d - z,\, V_1 = \frac{1}{2}e_1^2 \tag{13}$$

If the term k_1e_1 is added and subtracted to the $\dot{V}_1$ function, where $k_1 > 0$, it yields

$$\dot{V}_1 = e_1\dot{e}_1 = e_1(\dot{z}_d - V_z + k_1e_1 - k_1e_1) \tag{14}$$

$$\dot{V}_1 = -k_1e_1^2 + e_1(\dot{z}_d - V_z + k_1e_1) \tag{15}$$

The term $\dot{z}_d - v_z + k_1e_1$ of the Lyapunov function must vanish for a negative definite derivative, which can be achieved by choosing the virtual control v_z such that

$$v_{z_d} = \dot{z}_d + k_1e_1 + c_1\int e_1dt \tag{16}$$

Similar steps are repeated here to derive the control law,

$$e_2 = v_{z_d} - v_z,\, V_2 = \frac{1}{2}e_2^2 \tag{17}$$

Using a similar strategy as for v_{z_d} results

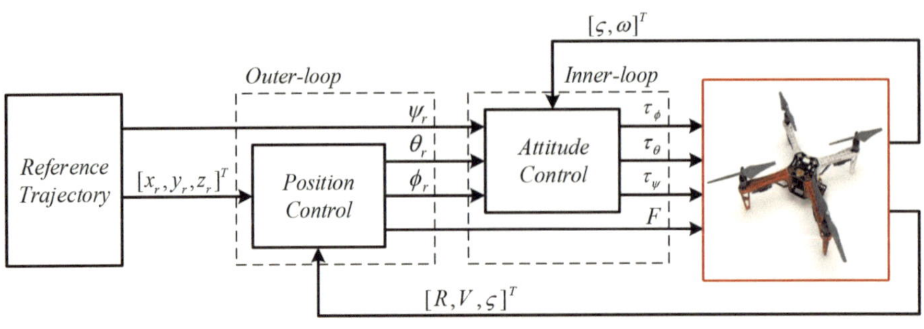

Figure 2.
The block diagram of the position control system of the quadcopter.

$$\dot{V}_2 = -k_2 e_2^2 + e_2 \left(\dot{v}_{z_d} - \frac{c\varphi c\theta}{m} U_1 + g + k_2 e_2 \right) \tag{18}$$

$$U_1 = \frac{m}{c\varphi c\theta} \{ \ddot{z}_d + k_1 \dot{e}_1 + g + k_2 e_2 \} \tag{19}$$

4.1.2 Attitude control

The control laws of the attitude of the quadcopter were derived in this section depending on integral backstepping method as follows:

$$U_2 = \frac{1}{b_1} \{ \ddot{\varphi}_d + k_3 \dot{e}_3 - a_1 \dot{\theta} \dot{\psi} - a_2 \dot{\theta} \Omega_r + k_4 e_4 \} \tag{20}$$

$$U_3 = \frac{1}{b_2} \{ \ddot{\theta}_d + k_5 \dot{e}_5 - a_2 \dot{\theta} \dot{\psi} + a_4 \dot{\theta} \Omega_r \} \tag{21}$$

$$U_4 = \frac{1}{b_3} \{ \ddot{\psi}_d + k_7 \dot{e}_7 - a_5 \dot{\varphi} \dot{\theta} \} \tag{22}$$

4.1.3 Position control

The Cartesian motion of a quadcopter in the x-y coordinate relies on θ and ϕ angles with respect to x and y axes, respectively. Hence, θ and ϕ angles have been considered as the outputs of x and y control laws. In this chapter, exact Euler angles, but not small Euler angles, have been considered to obtain the position control laws on x and y axes. However, this is an important criterion for high dynamic performance trajectory tracking control. The position control laws are derived from the quadcopter's model directly by applying the procedure of the control approaches. By applying the procedure of integral backstepping on the position equations of the quadcopter, one can obtain the following control laws:

$$\theta_d = \arcsin \left(\frac{m}{c\varphi . c\theta . U_1} \left\{ \dot{v}_{xd} - \frac{s\varphi . s\psi}{m} U_1 + k_{10} e_{10} \right\} \right) \tag{23}$$

$$\varphi_d = -\arcsin \left(\frac{m}{c\psi . U_1} \left\{ \dot{v}_{yd} - \frac{c\varphi . s\theta . s\psi}{m} U_1 + k_{12} e_{12} \right\} \right) \tag{24}$$

4.2 Quadcopter control using feedback linearization with LQI

The feedback linearization method is used in order to decouple the state variables of the quadcopter. This will enable us to derive the LQ-based control laws for the attitude, altitude, and position of the quadcopter.

4.2.1 Altitude control

The feedback linearization law of the attitude is given as follows:

$$Y_3 = z \tag{25}$$

$$\ddot{Y}_3 = \ddot{z} = \frac{c\varphi . c\theta}{m} U_1 - g \tag{26}$$

$$U_1 = \frac{m}{c\varphi . c\theta} (V_1 + g) \tag{27}$$

where V_1 is a virtual input, which is computed using LQI controller that will be presented in section 4.2.4.

4.2.2 Attitude control

The feedback linearization laws of the attitude are derived as follows:

$$U_2 = \frac{I_{xx}}{l}\{-a_1\dot{\theta}\dot{\psi} - a_2\Omega_r\dot{\theta} + V_2\} \tag{28}$$

$$U_3 = \frac{I_{yy}}{l}\{-a_3\dot{\varphi}\dot{\psi} + a_4\Omega_r\dot{\varphi} + V_3\} \tag{29}$$

$$U_4 = \frac{I_{zz}}{l}\{-a_5\dot{\varphi}\dot{\theta} + V_4\} \tag{30}$$

The previous control laws linearize the mapping between the derivatives of the flat outputs $Y_4 = \varphi$, $Y_5 = \theta$, $Y_6 = \psi$, and the virtual controls V_2, V_3, V_4. The latter are again computed using an LQI optimal controller,
where $a_1 = \frac{(I_{yy}-I_{zz})}{I_{xx}}$, $a_2 = \frac{J_r}{I_{xx}}$, $a_3 = \frac{(I_{zz}-I_{xx})}{I_{yy}}$, $a_4 = \frac{J_r}{I_{yy}}$, and $a_5 = \frac{(I_{zz}-I_{xx})}{I_{yy}}$.

4.2.3 Position control

Here, ϕ and θ angles are computed by the control laws of x and y motion, as it is done in the integral backstepping approach. The control laws are obtained as follows:

$$\theta_d = \arcsin\left(\frac{m}{c\varphi.c\theta.U_1}\left\{\dot{v}_{xd} - \frac{s\varphi.s\psi}{m}U_1 + V_5\right\}\right) \tag{31}$$

$$\varphi_d = -\arcsin\left(\frac{m}{c\psi.U_1}\left\{\dot{v}_{yd} - \frac{c\varphi.s\theta.s\psi}{m}U_1 + V_6\right\}\right) \tag{32}$$

where V_5 and V_6 are the proposed linear quadratic integral optimal controller.

4.2.4 Linear quadratic integral optimal control

The goal of the optimal control is to determine the control feedback, for which the optimal controller minimizes a proposed cost function J to desired minimum value. The cost function of the linear quadratic regulator is given as follows [32]:

$$J = \int_0^\infty (x^T Q\, x + u^T R\, u)dt \tag{33}$$

where Q and R represent the weighting matrices for the state vector x and control law vector u, respectively. LQR is conveniently applied to linear control systems or linearized nonlinear control systems. The state space model of a linear control system is given as follows:

$$\begin{aligned} \dot{x} &= Ax + Bu \\ y &= Cx + Du \end{aligned} \tag{34}$$

The control law u, which minimizes the cost function J, can be derived as follows:

$$u = -K\,x = -R^{-1}B^T P\,x \tag{35}$$

where P is a covariance matrix. It is the solution of the algebraic Riccati Eq. (36), in which $\dot{P} = 0$

$$A^T P + PA - PBR^{-1}B^T P + Q = \dot{P} \tag{36}$$

LQR controller is capable to provide a high dynamic performance when used with linear or linearized control systems. However, LQR is not capable to ensure fast tracking of time varying command signals [33, 34]. Different types of LQRs are demonstrated in literatures [32]. **Figure 3** shows an LQI regulator, with an integral action.

If the model of the linear system is extended by an error vector $\dot{\overline{z}}$ such as

$$\dot{\overline{z}} = r - y = r - (Cx + Du) \tag{37}$$

where r is a reference signal, which may represent the desired trajectory for tracking. The extended state space model of the LQI regulator is as follows:

$$\begin{bmatrix} \dot{\overline{x}} \\ \dot{\overline{z}} \end{bmatrix} = \begin{bmatrix} A & 0 \\ -C & 0 \end{bmatrix} \begin{bmatrix} \overline{x} \\ \overline{z} \end{bmatrix} + \begin{bmatrix} B & 0 \\ D & I \end{bmatrix} \begin{bmatrix} \overline{u} \\ r \end{bmatrix} \tag{38}$$

Hence, the control law $\overline{u}$ with an integral action is as follows:

$$\overline{u} = -K\,\overline{x} - K_I \overline{z} \tag{39}$$

4.3 Quadcopter control using sliding mode

In order to obtain the attitude and position control laws of the quadcopter using sliding mode control, the steps followed are discussed below.

4.3.1 Altitude control

In order to obtain the control laws of the quadcopter using sliding mode control, at first, the sliding surface should be determined as follows:

$$s_1 = c_1 e_1 + \dot{e}_1 \tag{40}$$

where $e_1 = z_d - z, \;\; \dot{e}_1 = \dot{z}_d - \dot{z}$, so that the derivative of the sliding surface becomes

$$\dot{s}_1 = c_1 \dot{e}_1 + \ddot{e}_1 \tag{41}$$

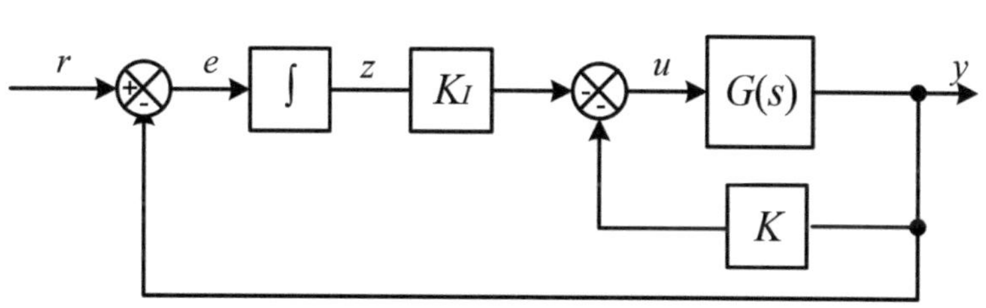

Figure 3.
LQI optimal controller structure.

From the equation of motion, the second derivative of the error becomes

$$\ddot{e}_1 = \ddot{z}_d - \ddot{z} = \ddot{z}_d - \frac{c\varphi c\theta}{m} U_1 + g \tag{42}$$

By equaling Eq. (41) to zero, we obtain

$$\dot{s}_1 = \ddot{z}_d - \frac{c\varphi c\theta}{m} U_1 + g + c_1(\dot{z}_d - \dot{z}) = 0 \tag{43}$$

By using the constant and proportional rate reaching law formula

$$-K_1 s_1 - Q_1 \operatorname{sgn}(s_1) = \ddot{z}_d - \frac{c\varphi c\theta}{m} U_1 + g + c_1(\dot{z}_d - \dot{z}) \tag{44}$$

So that the control law of the altitude will become:

$$U_1 = \frac{m}{c\varphi c\theta}\{\ddot{z}_d + g + c_1(\dot{z}_d - \dot{z}) + K_1 s_1 + Q_1 \operatorname{sgn}(s_1)\} \tag{45}$$

4.3.2 Attitude control

By following sliding mode control steps of design for the attitude of the quadcopter, we obtain

$$U_2 = \frac{1}{b_1}\{\ddot{\varphi}_d - a_1\dot{\theta}\dot{\psi} - a_2\dot{\theta}\Omega_r + c_2(\dot{\varphi}_d - \dot{\varphi}) + K_2 s_2 + Q_2.\operatorname{sgn}(s_2)\} \tag{46}$$

$$U_3 = \frac{1}{b_2}\{\ddot{\theta}_d - a_2\dot{\theta}\dot{\psi} + a_4\dot{\theta}\Omega_r + c_3(\dot{\theta}_d - \dot{\theta}) + K_3 s_3 + Q_3.\operatorname{sgn}(s_3)\} \tag{47}$$

$$U_4 = \frac{1}{b_3}\{\ddot{\psi}_d - a_5\dot{\varphi}\dot{\theta} + c_4(\dot{\psi}_d - \dot{\psi}) + K_4 s_4 + Q_4.\operatorname{sgn}(s_4)\} \tag{48}$$

with

$$s_2 = c_2 e_2 + \dot{e}_2 \quad s_3 = c_3 e_3 + \dot{e}_3 \quad s_4 = c_4 e_4 + \dot{e}_4$$

$$e_2 = \varphi_d - \varphi \quad e_3 = \theta_d - \theta \quad e_4 = \psi_d - \psi$$

4.3.3 Position control

Same strategy will be followed to derive the control laws of the position as in integral backstepping and feedback linearization. The control laws of both x, y will command the attitude loop with the references to accomplish the desired trajectory

$$\theta_d = \arcsin\left(\frac{m}{c\varphi.c\theta.U_1}\left\{\ddot{x}_d - \frac{s\varphi.s\psi}{m}U_1 + K_5 s_5 + Q_5.\operatorname{sgn}(s_5)\right\}\right) \tag{49}$$

$$\varphi_d = -\arcsin\left(\frac{m}{c\psi.U_1}\left\{\ddot{y}_d - \frac{c\varphi.s\theta.s\psi}{m}U_1 + K_6 s_6 + Q_6.\operatorname{sgn}(s_6)\right\}\right) \tag{50}$$

with

$$s_5 = c_5 e_5 + \dot{e}_5 \quad s_6 = c_6 e_6 + \dot{e}_6$$

$$e_5 = x_d - x \quad e_6 = y_d - y$$

DOI: http://dx.doi.org/10.5772/intechopen.86353

4.4 Results

The discussed nonlinear approaches have been tested in MATLAB/Simulink based on the nonlinear quadcopter model of Eq. (10), as well as experimental verification is also conducted. For modeling and simulation of the proposed approaches, the simulation sample time was Ts = 100 μs and the solver used was Runge-Kutta with a fixed integration. **Figures 4** and **5** show the system's trajectory tracking response. **Figure 4a** depicts the system response when implementing the proposed integral backstepping approach. **Figure 4b** shows the system response using feedback linearization with LQI approach. **Figure 4c** represents the system response using sliding mode control. **Figure 4a–c** demonstrates the system trajectory tracking to a desired trajectory command signal, with the existing external disturbances. These disturbances are being added with the command signals at different time instances. The initial position of the desired trajectory was (2, 0, 0), but the quadcopter was initiated with a different initial position as (0, 0, 0). As seen from **Figure 4a–c**, for the three investigated control approaches, the actual trajectory at the start was a bit diverged from the desired trajectory. However, the actual trajectory was then converged to the desired one fast. **Figure 5** exhibits the reference signals and the responses for x-, y-, and z axes of the quadcopter in the 3D space. These references on x and y axes were selected to be sinusoidal signals with 2 m of magnitude and 0.05 Hz of frequency. The command along z axis was a ramp signal with 0.2 $m.s^{-1}$ velocity rate. **Figure 6** shows the tracking errors of the

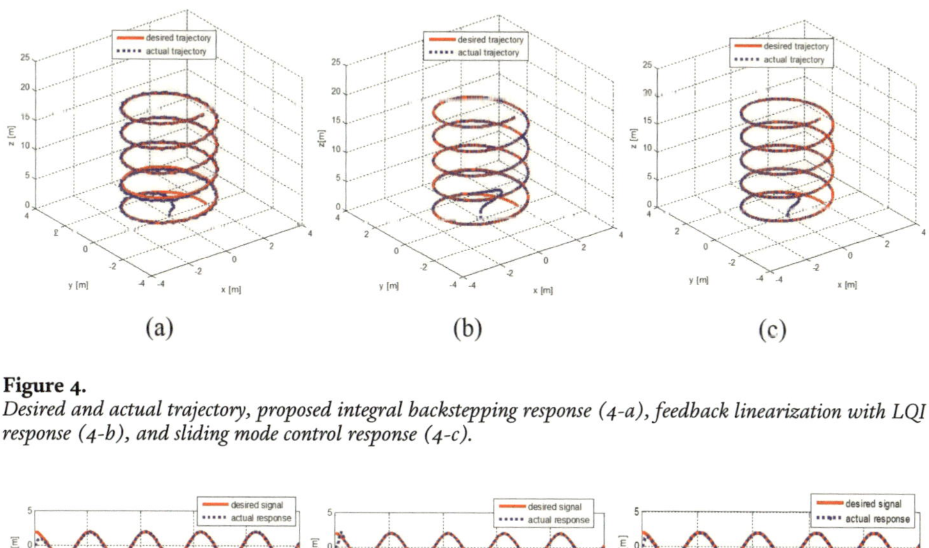

(a) (b) (c)

Figure 4.
Desired and actual trajectory, proposed integral backstepping response (4-a), feedback linearization with LQI response (4-b), and sliding mode control response (4-c).

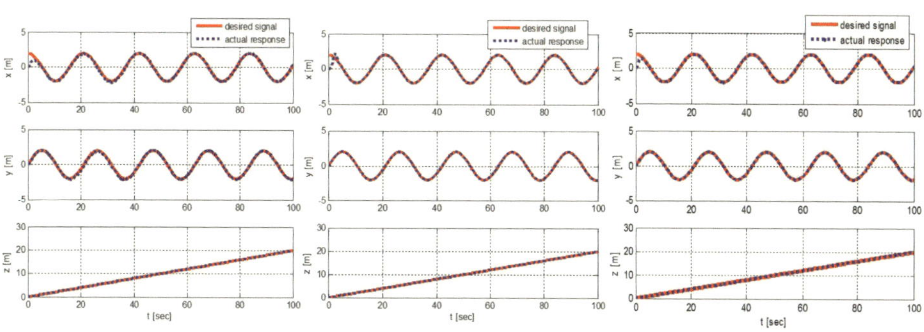

(a) (b) (c)

Figure 5.
Desired and actual trajectory, proposed integral backstepping response (5-a), feedback linearization with LQI response (5-b), and sliding mode control response (5-c).

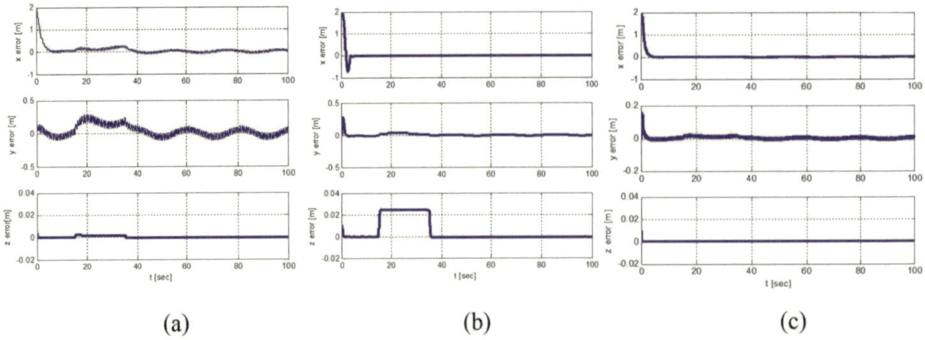

Figure 6.
Trajectory tracking errors, proposed integral backstepping response (6-a), feedback linearization with LQI response (6-b), and sliding mode control response (6-c).

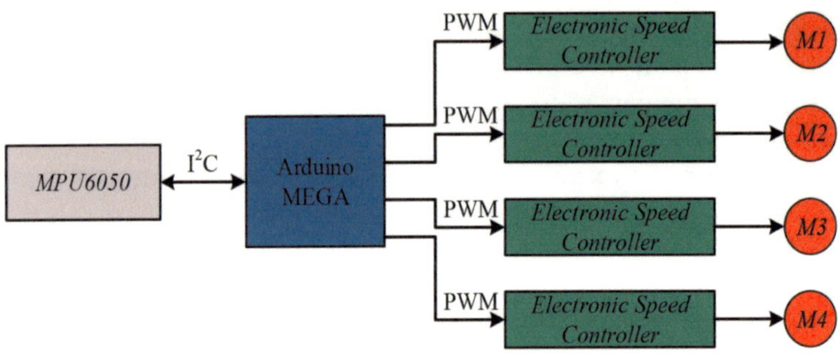

Figure 7.
Practical UAV control scheme.

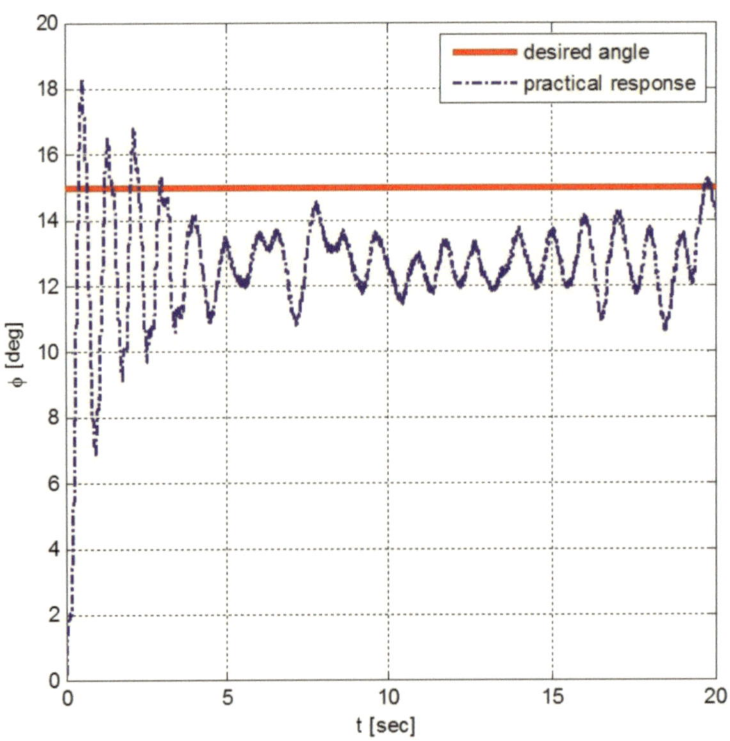

Figure 8.
Pitch practical response using integral backstepping.

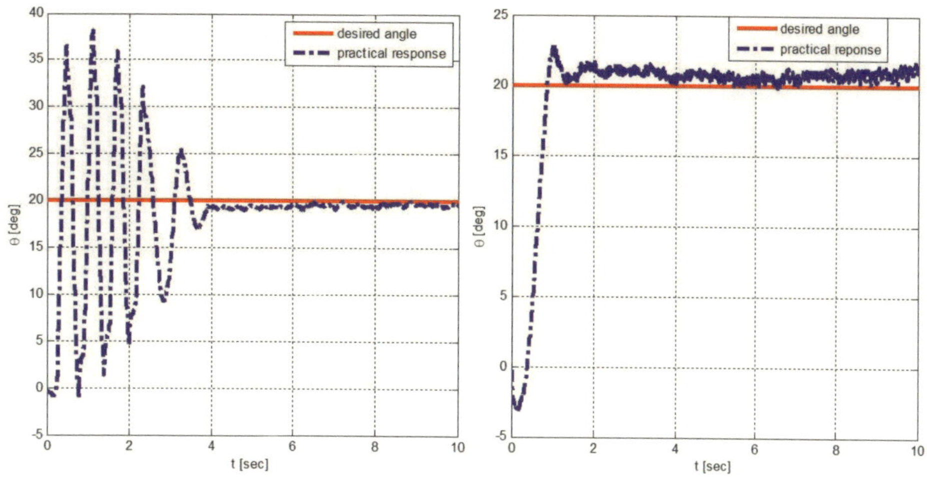

(a) (b)

Figure 9.
Pitch practical response: (a) using integral backstepping and (b) feedback linearization with LQI.

quadcopter motion on x, y and z. However, as seen, the tracking error of the motion on the three axes converged to zero. But, a little divergence was observed, which were due to the existence of disturbance with the command signals.

The practical implementation, of the proposed control strategies of the attitude control of the quadcopter, has been validated using Arduino MEGA board with an inertial measurement unit (IMU). **Figure 7** exhibits the practical UAV control system. **Figure 8** shows the practical implementation results and response of pitch angle using integral backstepping controller. As noticed earlier, there is a static error with oscillating response. **Figure 9a** and **b** demonstrates the practical result and response of the roll angle when implementing integral backstepping and feedback linearization with LQI controllers, respectively.

As noticed from **Figure 9a**, there was an oscillating response for pitch angle control during transient state, of almost undesired of $\pm 20°$ of overshoot and downshoot when implementing the proposed backstepping controller. But, high dynamic performance and fast tracking control were obtained for pitch angle control when implementing the proposed LQI controller with feedback linearization approach as seen in **Figure 9b**.

5. Conclusion

This chapter has discussed different advanced control techniques for UAV control. Nonlinear control theories have been reviewed among other control strategies due to their capacity to deal with the nonlinearity and the coupling components of the UAV state variables. This includes backstepping, feedback linearization, and sliding mode control. UAV nonlinear model has been derived and modeled in MATLAB®, and the proposed control strategies have been implemented. Simulation results obtained from the developed model with the control strategies were presented and discussed. Different path tracking and trajectories have been examined with successful and high dynamic performance. The developed control strategies have exhibited robustness against the UAV parameter mismatch and dynamic uncertainties.

Author details

Abdulkader Joukhadar[1*], Mohammad Alchehabi[2] and Adnan Jejeh[1]

1 Department of Mechatronics Engineering, University of Aleppo, Syria

2 Department of Control Engineering and Automation, University of Aleppo, Syria

*Address all correspondence to: ajoukhadar@alepuniv.edu.sy

References

[1] Ding X, Guo P, Xua K, Yu Y. A review of aerial manipulation of small-scale rotorcraft unmanned robotic systems. Chinese Journal of Aeronautics. 2018;**1069**:25

[2] Ruggiero F, Lippiello V, Ollero A. Aerial manipulation: A literature review. IEEE Robotics and Automation Letters. Jul 2018;**3**(3):1957-1964

[3] Cai F, Lai T, Chai Q, Wang W. Trajectory tracking problem of a quadrotor pendulum.In: 2016 28th Chinese Control and Decision Conference (CCDC). pp. 578-582

[4] Zhou S, Kang Y, Shi X. Indoor localization, navigation and mapping for quadrotor. In: Proceedings of 2014 IEEE Chinese Guidance, Navigation and Control Conference; 8–10 August 2014; Yantai. China. pp. 2669-2674

[5] Qian GM, Pebrianti D, Chun YW, Hao YH, Bayuaji L. Waypoint navigation of quadrotor MAV. In: 2017 7th IEEE International Conference on System Engineering and Technology (ICSET 2017); 2–3 October 2017; Shah Alam. Malaysia: pp. 38-42

[6] Guanglei M, Haibing P. The application of ultrasonic sensor in the obstacle avoidance of quadrotor UAV. In: Proceedings of 2016 IEEE Chinese Guidance, Navigation and Control Conference; 12–14 August 2016; Nanjing. China: pp. 976-981

[7] Zhao J, Li Y, Hu D, Pei Z. Design on altitude control system of quad rotor based on laser radar. In: 2016 IEEE/ CSAA International Conference on Aircraft Utility Systems (AUS); 10–12 October 2016; Beijing: China; 2016. pp. 105-109

[8] Ali Q, Montenegro S. Explicit model following distributed control scheme for formation flying of mini UAVs. IEEE Access. 2016;**4**:397-406

[9] Yamamoto K, Sekiguchi K, Nonaka K. Experimental verification of formation control by model predictive control considering collision avoidance in three dimensional space with quadcopter. In: 2017 11th Asian Control Conference (ASCC), Gold Coast Convention Centre; 17–20 December 2017; Australia. pp. 1602-1607

[10] Lin Q et al. Adaptive flight control design for quadrotor UAV based on dynamic inversion and Neural Networks. In: 2013 Third International Conference on Instrumentation, Measurement, Computer, Communication and Control. pp. 1461-1466

[11] Ding R, Jiao R, Chou W. Adaptive robust control for quad-rotor based on a compound control method. In: Proceedings of the 36th Chinese Control Conference; 26–28 July 2017; Dalian. China. pp. 4887-4893

[12] Song Z, Wang J. Adaptive trajectory tracking control of a quadrotor system based on fuzzy monitoring strategy. In: 8th International Conference on Information Science and Technology; June 30–July 6 2018; Granada, Cordoba, and Seville. Spain. pp. 415-421

[13] Wang Y, Yang P, Shu Q. Adaptive sliding mode control for distributed control systems with mismatched uncertainty. In: The 30th Chinese Control and Decision Conference (2018 CCDC). pp. 4673-4678

[14] Sun MP, Liu JJ. Tracking control of a quadrotor UAV based on T_S fuzzy model. In: Proceedings of the 36th Chinese Control Conference; 26–28 July 2017; Dalian. China. pp. 4216-4221

[15] Hwang CL. Hybrid neural network under-actuated sliding mode control for

trajectory tracking of quadrotor unmanned aerial vehicle. In: WCCI 2012 IEEE World Congress on Computational Intelligence; 10–15 June 2012; Brisbane. Australia. pp. 1-8

[16] Cisneros PSG, Hoffmann C, Bartels M, Werner H. Linear parameter varying controller design for a nonlinear quad rotor helicopter model for high speed trajectory tracking. In: 2016 American Control Conference (ACC), Boston Marriott Copley Place; 6–8 July 2016; Boston, MA, USA. pp. 486-491

[17] Tanveer MH et al. NMPC PID based control structure design for avoiding uncertainties in attitude and altitude tracking control of quadrotor (UAV). In: 2014 IEEE 10th International Colloquium on Signal Processing & its Applications (CSPA2014); 7–9 March 2014; Kuala Lumpur, Malaysia. pp. 117-122

[18] Alexis K, Nikolakopoulos G, Tzes A. Model predictive quadrotor control attitude, altitude and position experimental studies. IET Control Theory and Applications. 2012;**6**(12): 1812-1827

[19] Kumar R et al. Differential flatness based hybrid PID_LQR flight controller for complex trajectory tracking in quadcopter UAVs. IEEE. 2017:113-118

[20] Kuantama E, Tarca I, Tarca R. Feedback linearization LQR control for quadcopter position tracking. In: 2018 5th International Conference on Control, Decision and Information Technologies (CoDIT). 2018. pp. 204-209

[21] Joukhadar A, Alchehabi M, Muller A, Stroger C. Trajectory tracking control of a quadcopter UAV using nonlinear control approaches. In: 1st ICAMMRMS, Mechanism, Machine, Robotics and Mechatronics Sciences. Mechanisms and Machine Science. Vol. 58. Cham: Springer. pp. 271-285

[22] Li N, Yu S, Xi Z. Nonlinear control design for a quad rotor unmanned aerial vehicle. In: Proceedings of the 35th Chinese Control Conference; 27–29 July 2016; Chengdu. China. pp. 469-474

[23] Mellinger D, Kumar V. Minimum snap trajectory generation and control for quadrotors. In: IEEE International Conference on Robotics and Automation, IEEE; May 2011. pp. 2520-2525

[24] Falcón R et al. Quad rotor robust tracking: A continuous sliding mode control strategy. In: 2018 15th International Workshop on Variable Structure Systems (VSS); 9–11 July 2018, Graz University of Technology, Graz. Austria. pp. 390-395

[25] Yanning G, Zhaosen D, Liye Z, Yueyong L. Trajectory tracking control of a quadrotor using fractional order sliding mode. In: Proceedings of the 36th Chinese Control Conference; 26–28 July 2017; Dalian; China. pp. 6414–6419

[26] Khalil H. Nonlinear Systems. Dalian, USA: Prentice Hall; 2017. pp. 505-604

[27] Krstić M, Kanellakopoulos I, Kokotović P. Nonlinear Adaptive Control Design. 1st ed. New York: John Wiley and Sons; 1995. pp. 19-86

[28] Bandyopadhyay B, Janardhanan S. Discrete Time Sliding Mode Control. Springer; 2006. pp. 1-16

[29] Carillo L, López A, Lozano R, et al. Quad Rotorcraft Control. Springer; 2013. pp. 1-58

[30] Castillo P, Lozano R, Dzul AE. Modeling and Control of Mini-Flying Machines. London: Springer; 2005. pp. 1-56

[31] Bouabdullah S. Design and control of quadrotors with application to

autonomous flying [PhD thesis No. 3727], EPFL. 2007

[32] Anderson B, Moore J. Optimal Control: Linear Quadratic Method. New York: Prentice-Hall; 1989. pp. 1-99

[33] Joukhadar A, Hasan I, Alsabbagh A, Alkouzbary M. Integral Lqr-based 6dof autonomous quadrocopter balancing system control. International Journal of Advanced Research in Artificial Intelligence (IJARAI). 2015;**4**(5):10-17

[34] Alhaddad M, Shaukifeh B, Joukhadar A. Adaptive LQ-based computed-torque controller for robotic manipulator. IEEE Conference of Russian Young Researchers in Electrical and Electronic Engineering (EIConRus). Saint Petersburg and Moscow, Russia; 2019. pp. 2169-2173